D1051406

ONE
TWO
THREE . . .
INFINITY

There was a young fellow from Trinity
Who took $\sqrt{\infty}$
But the number of digits
Gave him the fidgets;
He dropped Math and took up Divinity.

ONE
TWO
THREE . . .
INFINITY

Facts and Speculations of Science

George Gamow

ILLUSTRATED BY THE AUTHOR

DOVER PUBLICATIONS, INC., *New York*

TO MY SON IGOR

WHO WANTED TO BE A COWBOY

This Dover edition, first published in 1988, is an unabridged and unaltered republication of the revised edition (1961) of the work first published by The Viking Press, Inc., N.Y., in 1947.

Manufactured in the United States of America
Dover Publications, Inc., 31 East 2nd Street, Mineola, N.Y. 11501

Library of Congress Cataloging-in-Publication Data

Gamow, George, 1904–1968.
   One, two, three—infinity.

   Reprint. Originally published: New York : Viking Press, 1961.
   Includes index.
   1. Science—Popular works.   I. Title.
Q162.G23   1988   500   88-18955
ISBN 0-486-25664-2 (pbk.)

Acknowledgment is made to Prentice-Hall, Inc. for permission to reproduce the drawings on pages 136–37, from *Matter, Earth and Sky* by George Gamow. © 1958, Prentice-Hall, Inc., Englewood Cliffs, N.J.

> "The time has come," the Walrus said,
> "To talk of many things"...
> Lewis Carroll, *Through the Looking-Glass*

Preface

... of atoms, stars, and nebulae, of entropy and genes; and whether one can bend space, and why the rocket shrinks. And indeed, in the course of this book we are going to discuss all these topics, and also many others of equal interest.

The book originated as an attempt to collect the most interesting facts and theories of modern science in such a way as to give the reader a general picture of the universe in its microscopic and macroscopic manifestations, as it presents itself to the eye of the scientist of today. In carrying out this broad plan, I have made no attempt to tell the whole story, knowing that any such attempt would inevitably result in an encyclopedia of many volumes. At the same time the subjects to be discussed have been selected so as to survey briefly the entire field of basic scientific knowledge, leaving no corner untouched.

Selection of subjects according to their importance and degree of interest, rather than according to their simplicity, necessarily has resulted in a certain unevenness of presentation. Some chapters of the book are simple enough to be understood by a child, whereas others will require some little concentration and study to be completely understood. It is hoped, however, that the layman reader will not encounter too serious difficulties in reading the book.

It will be noticed that the last part of the book, which discusses the "Macrocosmos," is considerably shorter than the part on "Microcosmos." This is primarily because I have already discussed in detail so many problems pertaining to the macrocosmos in *The Birth and Death of the Sun*, and *Biography of the Earth*,[1] and further detailed discussion here would be a tedious repeti-

[1] The Viking Press, New York, 1940 and 1941, respectively.

v

tion. Therefore in this part I have restricted myself to a general account of physical facts and events in the world of planets, stars, and nebulae and the laws that govern them, going into greater detail only in discussing problems upon which new light has been shed by the advance of scientific knowledge during the last few years. Following this principle I have given especial attention to the recent views according to which vast stellar explosions, known as "supernovae," are caused by the so-called "neutrinos," the smallest particles known in physics, and the new planetary theory, which abolishes the currently accepted views that planets originated as the result of collisions between the sun and some other stars, and re-establishes the old half-forgotten views of Kant and Laplace.

I want to express my thanks to numerous artists and illustrators whose work, topologically transformed (see Section II, Ch. III), has served as the basis for many illustrations adorning the book. Above all my thanks are due to my young friend Marina von Neumann, who claims that she knows everything better than her famous father does, except, of course, mathematics, which she says she knows only equally well. After she had read in manuscript some of the chapters of the book, and told me about numerous things in it which she could not understand, I finally decided that this book is not for children as I had originally intended it to be.

G. GAMOW

December 1, 1946

Preface to the 1961 Edition

All books on science are apt to become out of date a few years after publication, especially in the case of those branches of science which undergo rapid development. In this sense, my book *One Two Three . . . Infinity*, first published thirteen years ago, is a lucky one. It was written just after a number of important scientific advances, which were included in the text, and in order to bring it up to date relatively few changes and additions were necessary.

One of the important advances was the successful release of atomic energy by means of thermonuclear reactions in the form of H-bomb explosions, and the slow but steady progress toward the controlled release of energy through thermonuclear processes. Since the principle of thermonuclear reactions and their application in astrophysics were described in Chapter XI of the first edition of this book, man's progress toward the same goal could be taken care of simply by adding new material at the end of Chapter VII.

Other changes involved the increase in the estimated age of our universe from two or three billion years to five or more billion years, and the revised astronomical distance scale resulting from explorations with the new 200-inch Hale telescope on Mount Palomar in California.

Recent progress in biochemistry necessitated re-drawing Figure 101 and changing the text pertaining to it, as well as adding new material at the end of Chapter IX concerning synthetic production of simple living organisms. In the first edition I wrote (p. 266): "Yes, we certainly have a transitional step between living and non-living matter, and when—perhaps in no far-distant future—some talented biochemist is able to synthesize a virus molecule from ordinary chemical elements, he will be justified in exclaiming: 'I have just put the breath of life into a piece of dead matter!'" Well, a few years ago this was actually done, or almost done, in California, and the reader will find a short account of this work at the end of Chapter IX.

And one more change: The first printing of my book was dedicated "To my son Igor, who wants to be a cowboy." Many of my readers wrote me asking if he actually became a cowboy. The answer is no; he is graduating this summer, having majored in biology, and plans to work in genetics.

G. GAMOW

University of Colorado
November 1960

Contents

Illustrations

ONE
TWO
THREE   .   .   .
INFINITY

PART I

Playing with Numbers

CHAPTER I

Big Numbers

1. HOW HIGH CAN YOU COUNT?

THERE is a story about two Hungarian aristocrats who
decided to play a game in which the one who calls the
largest number wins.

"Well," said one of them, "you name your number first."

After a few minutes of hard mental work the second aristocrat
finally named the largest number he could think of.

"Three," he said.

Now it was the turn of the first one to do the thinking, but
after a quarter of an hour he finally gave up.

"You've won," he agreed.

Of course these two Hungarian aristocrats do not represent a
very high degree of intelligence[1] and this story is probably just a
malicious slander, but such a conversation might actually have
taken place if the two men had been, not Hungarians, but Hotten-
tots. We have it indeed on the authority of African explorers that
many Hottentot tribes do not have in their vocabulary the names
for numbers larger than three. Ask a native down there how many
sons he has or how many enemies he has slain, and if the number
is more than three he will answer "many." Thus in the Hottentot
country in the art of counting fierce warriors would be beaten
by an American child of kindergarten age who could boast the
ability to count up to ten!

Nowadays we are quite accustomed to the idea that we can
write as big a number as we please—whether it is to represent
war expenditures in cents, or stellar distances in inches—by

[1] This statement can be supported by another story of the same collection
in which a group of Hungarian aristocrats lost their way hiking in the Alps.
One of them, it is said, took out a map, and after studying it for a long
time, exclaimed: "Now I know where we are!" "Where?" asked the others.
"See that big mountain over there? We are right on top of it."

3

simply setting down a sufficient number of zeros on the right side of some figure. You can put in zeros until your hand gets tired, and before you know it you will have a number larger than even the total number of atoms in the universe,[2] which, incidentally, is 300,000,000,000,000,000,000,000,000,000,000,000,000,000,000,-000,000,000,000,000,000,000,000,000.

Or you may write it in this shorter form: $3 \cdot 10^{74}$.

Here the little number ^{74} above and to the right of 10 indicates that there must be that many zeros written out, or, in other words, 3 must be multiplied by 10 seventy-four times.

But this "arithmetic-made-easy" system was not known in ancient times. In fact it was invented less than two thousand years ago by some unknown Indian mathematician. Before his great discovery—and it *was* a great discovery, although we usually do not realize it—numbers were written by using a special symbol for each of what we now call decimal units, and repeating this symbol as many times as there were units. For example the number 8732 was written by ancient Egyptians:

𓏥𓏥𓏥𓏥𓏥𓏥𓏥𓏥 𓍢𓍢𓍢𓍢𓍢𓍢𓍢 𓎆𓎆𓎆

whereas a clerk in Caesar's office would have represented it in this form:

MMMMMMMMDCCXXXII

The latter notations must be familiar to you, since Roman numerals are still used sometimes—to indicate the volumes or chapters of a book, or to give the date of a historical event on a pompous memorial tablet. Since, however, the needs of ancient accounting did not exceed the numbers of a few thousands, the symbols for higher decimal units were nonexistent, and an ancient Roman, no matter how well trained in arithmetic, would have been extremely embarrassed if he had been asked to write "one million." The best he could have done to comply with the request, would have been to write one thousand *M's* in succession, which would have taken many hours of hard work (Figure 1).

For the ancients, very large numbers such as those of the stars

[2] Measured as far as the largest telescope can penetrate.

in the sky, the fish in the sea, or grains of sand on the beach were "incalculable," just as for a Hottentot "five" is incalculable, and becomes simply "many"!

It took the great brain of Archimedes, a celebrated scientist of the third century B.C., to show that it is possible to write really

FIGURE 1

An ancient Roman, resembling Augustus Caesar, tries to write "one million" in Roman numerals. All available space on the wall-board hardly suffices to write "a hundred thousand."

big numbers. In his treatise *The Psammites*, or *Sand Reckoner*, Archimedes says:

"There are some who think that the number of sand grains is infinite in multitude; and I mean by sand not only that which exists about Syracuse and the rest of Sicily, but all the grains of sand which may be found in all the regions of the Earth, whether inhabited or uninhabited. Again there are some who, without regarding the number as infinite, yet think that *no number can be named which is great enough to exceed that which would des-*

ignate the number of the Earth's grains of sand. And it is clear that those who hold this view, if they imagined a mass made up of sand in other respects as large as the mass of the Earth, including in it all the seas and all the hollows of the Earth filled up to the height of the highest mountains, would be still more certain that no number could be expressed which would be larger than that needed to represent the grains of sand thus accumulated. But I will try to show that of the numbers named by me some exceed not only the number of grains of sand which would make a mass equal in size to the Earth filled up in the way described, but even equal to a mass the size of the Universe."

The way to write very large numbers proposed by Archimedes in this famous work is similar to the way large numbers are written in modern science. He begins with the largest number that existed in ancient Greek arithmetic: a "myriad," or ten thousand. Then he introduced a new number, "a myriad myriad" (a hundred million), which he called "an octade" or a "unit of the second class." "Octade octades" (or ten million billions) is called a "unit of the third class," "octade, octade, octades" a "unit of the fourth class," etc.

The writing of large numbers may seem too trivial a matter to which to devote several pages of a book, but in the time of Archimedes the finding of a way to write big numbers was a great discovery and an important step forward in the science of mathematics.

To calculate the number representing the grains of sand necessary to fill up the entire universe, Archimedes had to know how big the universe was. In his time it was believed that the universe was enclosed by a crystal sphere to which the fixed stars were attached, and his famous contemporary Aristarchus of Samos, who was an astronomer, estimated the distance from the earth to the periphery of that celestial sphere as 10,000,000,000 stadia or about 1,000,000,000 miles.[3]

Comparing the size of that sphere with the size of a grain of sand, Archimedes completed a series of calculations that would give a highschool boy nightmares, and finally arrived at this conclusion:

[3] One Greek "stadium" is 606 ft. 6 in., or 188 meters (m).

"It is evident that the number of grains of sand that could be contained in a space as large as that bounded by the stellar sphere as estimated by Aristarchus, is not greater than one thousand myriads of units of the eighth class."[4]

It may be noticed here that Archimedes' estimate of the radius of the universe was rather less than that of modern scientists. The distance of one billion miles reaches only slightly beyond the planet Saturn of our solar system. As we shall see later the universe has now been explored with telescopes to the distance of 5,000,000,000,000,000,000,000 miles, so that the number of sand grains necessary to fill up all the visible universe would be over:

$$10^{100} \text{ (that is, 1 and 100 zeros)}$$

This is of course much larger than the total number of atoms in the universe, $3 \cdot 10^{74}$, as stated at the beginning of this chapter, but we must not forget that the universe is *not packed* with atoms; in fact there is on the average only about 1 atom per cubic meter of space.

But it isn't at all necessary to do such drastic things as packing the entire universe with sand in order to get really large numbers. In fact they very often pop up in what may seem at first sight a very simple problem, in which you would never expect to find any number larger than a few thousands.

One victim of overwhelming numbers was King Shirham of India, who, according to an old legend, wanted to reward his grand vizier Sissa Ben Dahir for inventing and presenting to him the game of chess. The desires of the clever vizier seemed very modest. "Majesty," he said kneeling in front of the king, "give me a grain of wheat to put on the first square of this chessboard, and two grains to put on the second square, and four grains to put on the third, and eight grains to put on the fourth. And so, oh King, doubling the number for each succeeding square, give me enough grains to cover all 64 squares of the board."

[4] In our notation it would be:

thousand myriads	2nd class	3rd class	4th class
(10,000,000) ×	(100,000,000) ×	(100,000,000) ×	(100,000,000) ×
5th class	6th class	7th class	8th class
(100,000,000) ×	(100,000,000) ×	(100,000,000) ×	(100,000,000)

or simply:

$$10^{63} \text{ (i.e., 1 and 63 zeros)}$$

"You do not ask for much, oh my faithful servant," exclaimed the king, silently enjoying the thought that his liberal proposal of a gift to the inventor of the miraculous game would not cost him much of his treasure. "Your wish will certainly be granted." And he ordered a bag of wheat to be brought to the throne.

But when the counting began, with 1 grain for the first square, 2 for the second, 4 for the third and so forth, the bag was emptied

FIGURE 2

Grand Vizier Sissa Ben Dahir, a skilled mathematician, asks his reward from King Shirham of India.

before the twentieth square was accounted for. More bags of wheat were brought before the king but the number of grains needed for each succeeding square increased so rapidly that it soon became clear that with all the crop of India the king could not fulfill his promise to Sissa Ben. To do so would have required 18,446,744,073,709,551,615 grains![5]

[5] The number of wheat grains that the clever vizier had demanded may be represented as follows:

$$1+2+2^2+2^3+2^4+ \ldots +2^{62}+2^{63}.$$

In arithmetic a sequence of numbers each of which is progressively increased by the same factor (in this case by a factor of 2) is known as *geometrical progression*. It can be shown that the sum of all the terms in such a progression may be found by raising the constant factor (in this case 2) to the power represented by the number of steps in the progression

That's not so large a number as the total number of atoms in the universe, but it is pretty big anyway. Assuming that a bushel of wheat contains about 5,000,000 grains, one would need some 4000 billion bushels to satisfy the demand of Sissa Ben. Since the world production of wheat averages about 2,000,000,000 bushels a year, the amount requested by the grand vizier was that of the *world's wheat production for the period of some two thousand years!*

Thus King Shirham found himself deep in debt to his vizier and had either to face the incessant flow of the latter's demands, or to cut his head off. We suspect that he chose the latter alternative.

Another story in which a large number plays the chief role also comes from India and pertains to the problem of the "End of the World." W. W. R. Ball, the historian of mathematical fancy, tells the story in the following words:[6]

In the great temple at Benares beneath the dome which marks the center of the world, rests a brass plate in which are fixed three diamond needles, each a cubit high (a cubit is about 20 inches) and as thick as the body of a bee. On one of these needles, at the creation, God placed sixty-four discs of pure gold, the largest disc resting on the brass plate and the others getting smaller and smaller up to the top one. This is the tower of Brahma. Day and night unceasingly, the priest on duty transfers the discs from one diamond needle to another, according to the fixed and immutable laws of Brahma, which require that the priest must move only one disc at a time, and he must place these discs on needles so that there never is a smaller disc below a larger one. When all the sixty-four discs shall have been thus transferred from the

---

(in this case, 64), subtracting the first term (in this case, 1), and dividing the result by the above-mentioned constant factor minus 1. It may be stated thus:

$$\frac{2^{63} \times 2 - 1}{2 - 1} = 2^{64} - 1$$

and writing it as an explicit number:

18,446,744,073,709,551,615.

[6] W. W. R. Ball, *Mathematical Recreations and Essays* (The Macmillan Co., New York, 1939).

needle on which, at the creation, God placed them, to one of the other needles, tower, temple, and Brahmans alike will crumble into dust, and with a thunderclap the world will vanish.

Figure 3 is a picture of the arrangement described in the story, except that it shows a smaller number of discs. You can make this puzzle toy yourself by using ordinary cardboard discs instead of golden ones, and long iron nails instead of the diamond needles

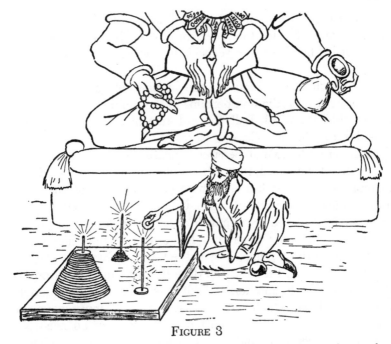

FIGURE 3

A priest working on the "End of the World" problem in front of a giant statue of Brahma. The number of golden discs is shown here smaller than 64 because it was difficult to draw so many.

of the Indian legend. It is not difficult to find the general rule according to which the discs have to be moved, and when you find it you will see that the transfer of each disc requires twice as many moves as that of the previous one. The first disc requires just one move, but the number of moves required for each succeeding disc increases geometrically, so that when the 64th disc is reached it must be moved as many times as there were grains in the amount of wheat Sissa Ben Dahir requested![7]

How long would it take to transfer all sixty-four discs in the tower of Brahma from one needle to the other? Suppose that priests worked day and night without holidays or vacation, making one move every second. Since a year contains about 31,558,000 seconds it would take slightly more than *fifty-eight thousand billion years* to accomplish the job.

It is interesting to compare this purely legendary prophecy of the duration of the universe with the prediction of modern science. According to the present theory concerning the evolution of the universe, the stars, the sun, and the planets, including our Earth, were formed about 3,000,000,000 years ago from shapeless masses. We also know that the "atomic fuel" that energizes the stars, and in particular our sun, can last for another 10,000,000,000 or 15,000,000,000 years. (See the chapter on "The Days of Creation.") Thus the total life period of our universe is definitely shorter than 20,000,000,000 years, rather than as long as the 58,000 billion years estimated by Indian legend! But, after all, it is only a legend!

Probably the largest number ever mentioned in literature pertains to the famous "Problem of a Printed Line." Suppose we built a printing press that would continuously print one line after another, automatically selecting for each line a different combination of the letters of the alphabet and other typographical signs. Such a machine would consist of a number of separate discs with the letters and signs all along the rim. The discs would be geared to one another in the same way as the number discs in

---

[7] If we have only seven discs the number of necessary moves is:

$$1+2^1+2^2+2^3+\text{etc., or}$$
$$2^7-1=2\cdot2\cdot2\cdot2\cdot2\cdot2\cdot2-1=127.$$

If you moved the discs rapidly without making any mistakes it would take you about an hour to complete the task. With 64 disks the total number of moves necessary is:

$$2^{64}-1=18,446,744,073,709,551,615$$

this is the same as the number of grains of wheat required by Sissa Ben Dahir.

the mileage indicator of your car, so that a full rotation of each disc would move the next one forward one place. The paper as it comes from a roll would automatically be pressed to the cylinder after each move. Such an automatic printing press could be built without much difficulty, and what it would look like is represented schematically in Figure 4.

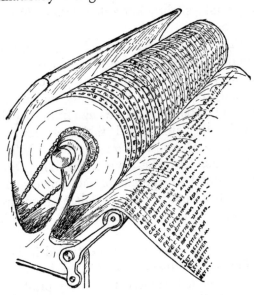

FIGURE 4

An automatic printing press that has just printed correctly a line from Shakespeare.

Let us set the machine in action and inspect the endless sequence of different printed lines that come from the press. Most of the lines make no sense at all. They look like this:

"aaaaaaaaaaa . . ."

or

"boobooboobooboo . . ."

or again:

"zawkporpkossscilm . . ."

But since the machine prints *all possible* combinations of letters and signs, we find among the senseless trash various sentences that have meaning. There are, of course, a lot of useless sentences such as:

"horse has six legs and . . ."

or

"I like apples cooked in terpentin. . . ."

But a search will reveal also every line written by Shakespeare, even those from the sheets that he himself threw into the wastepaper basket!

In fact such an automatic press would print everything that was ever written from the time people learned to write: every line of prose and poetry, every editorial and advertisement from newspapers, every ponderous volume of scientific treatises, every love letter, every note to a milkman. . . .

Moreover the machine would print everything that is to be printed in centuries to come. On the paper coming from the rotating cylinder we should find the poetry of the thirtieth century, scientific discoveries of the future, speeches to be made in the 500th Congress of the United States, and accounts of intraplanetary traffic accidents of the year 2344. There would be pages and pages of short stories and long novels, never yet written by human hand, and publishers having such machines in their basements would have only to select and edit good pieces from a lot of trash—which they are doing now anyway.

Why cannot this be done?

Well, let us count the number of lines that would be printed by the machine in order to present all possible combinations of letters and other typographical signs.

There are 26 letters in the English alphabet, ten figures (0, 1, 2 . . . 9) and 14 common signs (blank space, period, comma, colon, semicolon, question mark, exclamation mark, dash, hyphen, quotation mark, apostrophe, brackets, parentheses, braces); altogether 50 symbols. Let us also assume that the machine has 65 wheels corresponding to 65 places in an average printed line. The printed line can begin with any of these signs so that we have here 50 possibilities. For *each* of these 50 possibilities there are 50 possibilities for the second place in the line; that is, altogether $50 \times 50 = 2500$ possibilities. But for each given combination of the first two letters we have the choice between 50 possible signs in the third place, and so forth. Altogether the number of possible arrangements in the entire line may be expressed as:

$$\overbrace{50 \times 50 \times 50 \times \ldots \times 50}^{65 \text{ times}}$$

or 50^{65}

which is equal to 10^{110}

To feel the immensity of that number assume that each atom in the universe represents a separate printing press, so that we have $3 \cdot 10^{74}$ machines working simultaneously. Assume further that all these machines have been working continuously since the creation of the universe, that is for the period of 3 billion years or 10^{17} seconds, printing at the rate of atomic vibrations, that is, 10^{15} lines per second. By now they would have printed about

$$3 \cdot 10^{74} \times 10^{17} \times 10^{15} = 3 \cdot 10^{106}$$

lines—which is only about one thirtieth of 1 per cent of the total number required.

Yes, it would take a very long time indeed to make any kind of selection among all this automatically printed material!

2. HOW TO COUNT INFINITIES

In the preceding section we discussed numbers, many of them fairly large ones. But although such numerical giants as the number of grains of wheat demanded by Sissa Ben are almost unbelievably large, they are still finite and, given enough time, one could write them down to the last decimal.

But there are some really infinite numbers, which are larger than any number we can possibly write no matter how long we work. Thus "the number of all numbers" is clearly infinite, and so is "the number of all geometrical points on a line." Is there anything to be said about such numbers except that they are infinite, or is it possible, for example, to compare two different infinities and to see which one is "larger"?

Is there any sense in asking: "Is the number of all numbers larger or smaller than the number of all points on a line?" Such questions as this, which at first sight seem fantastic, were first considered by the famous mathematician Georg Cantor, who can be truly named the founder of the "arithmetics of infinity."

If we want to speak about larger and smaller infinities we face a problem of comparing the numbers that we can neither name

FIGURE 5

An African native and Prof. G. Cantor comparing the numbers beyond their counting ability.

nor write down, and are more or less in the position of a Hottentot inspecting his treasure chest and wanting to know whether he has more glass beads or more copper coins in his possession. But, as you will remember, the Hottentot is unable to count beyond three. Then shall he give up all attempts to compare the number of beads and the number of coins because he cannot count them? Not at all. If he is clever enough he will get his answer by comparing the beads and the coins piece by piece. He will place one bead near one coin, another bead near another coin, and so on, and so on . . . If he runs out of beads while there are still some coins, he knows that he has more coins than beads; if he runs out of coins with some beads left he knows that he has more beads than coins, and if he comes out even he knows that he has the same number of beads as coins.

Exactly the same method was proposed by Cantor for comparing two infinities: if we can pair the objects of two infinite groups so that each object of one infinite collection pairs with each object of another infinite collection, and no objects in either group are left alone, the two infinities are equal. If, however, such arrangement is impossible and in one of the collections some unpaired objects are left, we say that the infinity of objects in this collection is larger, or we can say stronger, than the infinity of objects in the other collection.

This is evidently the most reasonable, and as a matter of fact the only possible, rule that one can use to compare infinite quantities, but we must be prepared for some surprises when we actually begin to apply it. Take for example, the infinity of all even and the infinity of all odd numbers. You feel, of course, intuitively that there are as many even numbers as there are odd, and this is in complete agreement with the above rule, since a one-to-one correspondence of these numbers can be arranged:

$$1 \quad 3 \quad 5 \quad 7 \quad 9 \quad 11 \quad 13 \quad 15 \quad 17 \quad 19 \quad \text{etc.}$$
$$\updownarrow \quad \updownarrow \quad \updownarrow \quad \updownarrow \quad \updownarrow \quad \updownarrow \quad \updownarrow \quad \updownarrow \quad \updownarrow \quad \updownarrow$$
$$2 \quad 4 \quad 6 \quad 8 \quad 10 \quad 12 \quad 14 \quad 16 \quad 18 \quad 20 \quad \text{etc.}$$

There is an even number to correspond with each odd number in this table, and vice versa; hence the infinity of even numbers is equal to the infinity of odd numbers. Seems quite simple and natural indeed!

But wait a moment. Which do you think is larger: the number of all numbers, both even and odd, or the number of even numbers only? Of course you would say the number of all numbers is larger because it contains in itself all even numbers and in addition all odd ones. But that is just your impression, and in order to get the exact answer you must use the above rule for comparing two infinities. And if you use it you will find to your surprise that your impression was wrong. In fact here is the table of one-to-one correspondence of all numbers on one side, and even numbers only on the other:

$$1 \quad 2 \quad 3 \quad 4 \quad 5 \quad 6 \quad 7 \quad 8 \quad \text{etc.}$$
$$\updownarrow \quad \updownarrow \quad \updownarrow \quad \updownarrow \quad \updownarrow \quad \updownarrow \quad \updownarrow \quad \updownarrow$$
$$2 \quad 4 \quad 6 \quad 8 \quad 10 \quad 12 \quad 14 \quad 16 \quad \text{etc.}$$

According to our rule of comparing infinities we must say that the infinity of even numbers is exactly as large as the infinity of all numbers. This sounds, of course, paradoxical, since even numbers represent only a part of all numbers, but we must remember that we operate here with infinite numbers, and must be prepared to encounter different properties.

In fact in the world of infinity *a part may be equal to the whole!* This is probably best illustrated by an example taken from one of the stories about the famous German mathematician David Hilbert. They say that in his lectures on infinity he put this paradoxical property of infinite numbers in the following words:[8]

"Let us imagine a hotel with a finite number of rooms, and assume that all the rooms are occupied. A new guest arrives and asks for a room. 'Sorry—says the proprietor—but all the rooms are occupied.' Now let us imagine a hotel with an *infinite* number of rooms, and all the rooms are occupied. To this hotel, too, comes a new guest and asks for a room.

" 'But of course!'—exclaims the proprietor, and he moves the person previously occupying room N1 into room N2, the person from room N2 into room N3, the person from room N3 into room N4, and so on. . . . And the new customer receives room N1, which became free as the result of these transpositions.

"Let us imagine now a hotel with an infinite number of rooms, all taken up, and an infinite number of new guests who come in and ask for rooms.

" 'Certainly, gentlemen,' says the proprietor, 'just wait a minute.'

"He moves the occupant of N1 into N2, the occupant of N2 into N4, the occupant of N3 into N6, and so on, and so on . . .

[8] From the unpublished, and even never written, but widely circulating volume: "The Complete Collection of Hilbert Stories" by R. Courant.

"Now all odd-numbered rooms become free and the infinity or new guests can easily be accommodated in them."

Well, it is not easy to imagine the conditions described by Hilbert even in Washington as it was during the war, but this example certainly drives home the point that in operating with infinite numbers we encounter properties rather different from those to which we are accustomed in ordinary arithmetic.

Following Cantor's rule for comparing two infinities, we can also prove now that the number of all ordinary arithmetical fractions like $\frac{3}{7}$ or $\frac{735}{8}$ is the same as the number of all integers. In fact we can arrange all ordinary fractions in a row according to the following rule: Write first the fractions for which the sum of the numerator and denominator is equal to 2; there is only one such fraction namely: $\frac{1}{1}$. Then write fractions with sums equal to 3: $\frac{2}{1}$ and $\frac{1}{2}$. Then those with sums equal to 4: $\frac{3}{1}, \frac{2}{2}, \frac{1}{3}$. And so on. In following this procedure we shall get an infinite sequence of fractions, containing every single fraction one can think of (Figure 5). Now write above this sequence, the sequence of integers and you have the one-to-one correspondence between the infinity of fractions and the infinity of integers. Thus their number is the same!

"Well, it is all very nice," you may say, "but doesn't it mean simply that *all* infinities are equal to one another? And if that's the case, what's the use of comparing them anyway?"

No, that is not the case, and one can easily find the infinity that is larger than the infinity of all integers or all arithmetical fractions.

In fact, if we examine the question asked earlier in this chapter about the number of points on a line as compared with the number of all integer numbers, we find that these two infinities are different; there are many more points on a line than there are integers or fractional numbers. To prove this statement let us try to establish one-to-one correspondence between the points on a line, say 1 in. long, and the sequence of integer numbers.

Each point on the line is characterized by its distance from one end of the line, and this distance can be written in the form of an infinite decimal fraction, like 0.7350624780056 . . . . or 0.38250375632 . . . .[9] Thus we have to compare the number of all integers with the number of all possible infinite decimal fractions. What is the difference now between the infinite decimal fractions, as given above, and ordinary arithmetical fractions like $\frac{3}{7}$ or $\frac{8}{277}$?

You must remember from your arithmetic that every ordinary fraction can be converted into an infinite *periodic* decimal fraction. Thus $\frac{2}{3} = 0.66666$ . . . . $= 0.(6)$, and $\frac{3}{7} = 0.428571\,4\,28571\,4$ $28571\,4 . . . = 0.(428571)$. We have proved above that the number of all ordinary arithmetical fractions is the same as the number of all integers; so the number of all *periodic* decimal fractions must also be the same as the number of all integers. But the points on a line are not necessarily represented by *periodic* decimal fractions, and in most cases we shall get the infinite fractions in which the decimal figures appear without any periodicity at all. And it is easy to show that in such case no linear arrangement is possible.

Suppose that somebody claims to have made such an arrangement, and that it looks something like this:

N	
1	0.38602563078 . . . .
2	0.57350762050 . . . .
3	0.99356753207 . . . .
4	0.25763200456 . . . .
5	0.00005320562 . . . .
6	0.99035638567 . . . .
7	0.55522730567 . . . .
8	0.05277365642 . . . .
·	. . . .
·	. . . .
·	. . . .
·	. . . .
·	. . . .

[9] All these fractions are smaller than unity, since we have assumed the length of the line to be one.

Of course, since it is impossible actually to write the infinity of numbers with the infinite number of decimals in each, the above claim means that the author of the table has some general rule (similar to one used by us for arrangement of ordinary fractions) according to which he has constructed the table, and this rule guarantees that every single decimal fraction one can think of will appear sooner or later in the table.

Well, it is not at all difficult to show that any claim of that kind is unsound, since we can always write an infinite decimal fraction that is *not* contained in this infinite table. How can we do it? Oh, very simply. Just write the fraction with the first decimal different from that of N1 in the table, the second decimal different from that in N2 of the table and so on. The number you will get will look something like this:

$$0.\underset{\downarrow}{5}\;\underset{\downarrow}{2}\;\underset{\downarrow}{7}\;\underset{\downarrow}{4}\;\underset{\downarrow}{0}\;\underset{\downarrow}{7}\;\underset{\downarrow}{1}\;\underset{\downarrow}{2}\quad\text{etc.}$$

not 3 not 7 not 3 not 6 not 5 not 6 not 3 not 5

and this number is not included in the table no matter how far down you look for it. In fact if the author of the table will tell you that this very fraction you have written here stands under the No. 137 (or any other number) in his table you can answer immediately: "No, it isn't the same fraction because the one hundred and thirty seventh decimal in your fraction is different from the one hundred and thirty seventh decimal in the fraction I have in mind."

Thus it is impossible to establish a one-to-one correspondence between the points on a line and the integer numbers, which means that *the infinity of points on a line is larger, or stronger, than the infinity of all integer or fractional numbers.*

We have been discussing the points on a line "1 in. long," but it is easy to show now that, according to the rules of our "infinity arithmetics," the same is true of a line of any length. In fact, *there is the same number of points in lines one inch, one foot, or one mile long.* In order to prove it just look at Figure 6, which compares the number of points on two lines *AB* and *AC* of dif-

ferent lengths. To establish the one-to-one correspondence between the points of these two lines we draw through each point
on AB a line parallel to BC, and pair the points of intersections as
for example D and D^1, E and E^1, F and F^1, etc. Each point on AB
has a corresponding point on AC and vice versa; thus according
to our rule the two infinities of points are equal.

A still more striking result of the analysis of infinity consists in
the statement that: *the number of all points on a plane is equal
to the number of all points on a line.* To prove this let us consider
the points on a line AB one inch long, and the points within a
square $CDEF$ (Figure 7).

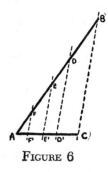

FIGURE 6

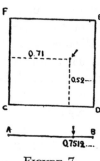

FIGURE 7

Suppose that the position of a certain point on the line is given
by some number, say 0.75120386 . . . . We can make from this
number two different numbers selecting even and odd decimal
signs and putting them together. We get this:

$$0.7108 \dots$$

and this:

$$0.5236 \dots$$

Measure the distances given by these numbers in the horizontal
and vertical direction in our square, and call the point so obtained
the "pair-point" to our original point on the line. In reverse, if we
have a point in the square the position of which is described by,
let us say, the numbers:

$$0.4835 \dots$$

and

$$0.9907 \dots$$

we obtain the position of the corresponding "pair-point" on the line by merging these two numbers:

$$0.49893057 \ldots$$

It is clear that this procedure establishes the one-to-one relationship between two sets of points. Every point on the line will have its pair in the square, every point in the square will have its pair on the line, and no points will be left over. Thus according to the criterion of Cantor, the infinity of all the points within a square is equal to the infinity of all the points on a line.

In a similar way it is easy to prove also that the infinity of all points within a cube is the same as the infinity of points within a square or on a line. To do this we merely have to break the original decimal fraction into three parts,[10] and use the three new fractions so obtained to define the position of the "pair-point" inside the cube. And, just as in the case of two lines of different lengths, the number of points within a square or a cube will be the same regardless of their size.

But the number of all geometrical points, though larger than the number of all integer and fractional numbers, is not the largest one known to mathematicians. In fact it was found that the *variety of all possible curves, including those of most unusual shapes, has a larger membership than the collection of all geometrical points, and thus has to be described by the third number of the infinite sequence.*

According to Georg Cantor, the creator of the "arithmetics of infinity," infinite numbers are denoted by the Hebrew letter $\aleph$ (aleph) with a little number in the lower right corner that indicates the order of the infinity. The sequence of numbers (including the infinite ones!) now runs:

$$1. \ 2. \ 3. \ 4. \ 5. \ldots \ldots \aleph_1 \ \aleph_2 \ \aleph_3 \ldots \ldots$$

and we say "there are $\aleph_1$ points on a line" or "there are $\aleph_2$

[10] For example from

$$0. \ 735106822548312 \ldots \ldots \text{etc.}$$

we make

$$0. \ 71853 \ldots$$
$$0. \ 30241 \ldots$$
$$0. \ 56282 \ldots$$

different curves," just as we say that "there are 7 parts of the world" or "52 cards in a pack."

In concluding our talk about infinite numbers we point out that these numbers very quickly outrun any thinkable collection to which they can possibly be applied. We know that $\aleph$ represents the number of all integers, $\aleph_1$ represents the number of all

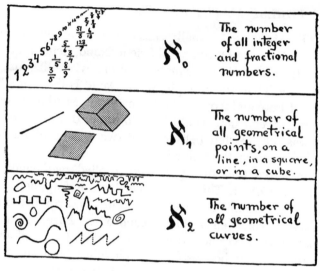

FIGURE 8

The first three infinite numbers.

geometrical points, and $\aleph_2$ the number of all curves, but nobody as yet has been able to conceive any definite infinite collection of objects that should be described by $\aleph_3$. It seems that the three first infinite numbers are enough to count anything we can think of, and we find ourselves here in a position exactly opposite to that of our old friend the Hottentot who had many sons but could not count beyond three!

CHAPTER II

Natural and Artificial Numbers

1. *THE PUREST MATHEMATICS*

MATHEMATICS is usually considered, especially by mathematicians, the Queen of all Sciences and, being a queen, it naturally tries to avoid morganatic relations with other branches of knowledge. Thus, for example, when David Hilbert, at a "Joint Congress of Pure and Applied Mathematics," was asked to deliver an opening speech that would help to break down the hostility that, it was felt, existed between the two groups of mathematicians, he began in the following way:

"We are often told that pure and applied mathematics are hostile to each other. This is not true. Pure and applied mathematics are not hostile to each other. Pure and applied mathematics have never been hostile to each other. Pure and applied mathematics will never be hostile to each other. Pure and applied mathematics cannot be hostile to each other because, in fact, there is absolutely nothing in common between them."

But although mathematics likes to be pure and to stand quite apart from other sciences, other sciences, especially physics, like mathematics, and try to "fraternize" with it as much as possible. In fact, almost every branch of pure mathematics is now being put to work to explain one or another feature of the physical universe. This includes such disciplines as the theory of abstract groups, noncommutable algebra, and non-Euclidian geometry, which have always been considered most pure and incapable of any application whatever.

One large system of mathematics, however, has up to now managed to remain quite useless for any purpose except that of stimulating mental gymnastics, and thus can carry with honor the "crown of purity." This is the so-called "theory of numbers" (meaning integer numbers), one of the oldest and most intricate products of pure mathematical thought.

Strange as it may seem, the theory of numbers, being the purest kind of mathematics, can be called, from a certain aspect, an empirical or even an experimental science. In fact most of its propositions have been formulated as a result of trying to do different things with numbers, in the same way that the laws of physics have resulted from trying to do different things with material objects. And just as in physics, some of these propositions have been proved "mathematically," whereas others still remain of purely empirical origin, and that are still challenging the brains of the best mathematicians.

Take, for example, the problem of prime numbers, that is, the numbers that cannot be represented as the product of two or more smaller numbers. 1, 2, 3, 5, 7, 11, 13, 17, etc. are such prime numbers, whereas 12, for example, is not, since it can be written as $2 \times 2 \times 3$.

Is the number of primes unlimited, or is there a largest prime beyond which each number can be represented as the product of the primes we already have? This problem was first attacked by Euclid himself, who gave a very simple and elegant proof that the number of primes extends beyond any limit so that there is no such thing as the "largest prime."

In order to examine this question suppose for a moment that there is known only a finite number of primes, and that some large number designated by the letter N represents the highest prime number known. Now let us get the product of all known primes, and add 1 to it. We can write it in this form:

$$(1 \times 2 \times 3 \times 5 \times 7 \times 11 \times 13 \times \ldots \times N) + 1.$$

It is of course much larger than the alleged "largest prime number" N. It is clear, however, that this number cannot be divided exactly by any of our primes (up to and including N) since from the way it is constructed we see that the division by any of these primes will leave the remainder 1.

Thus our number must be either a prime number itself, or must be divisible by a prime larger than N, both of which cases contradict our original assumption that N is the largest existing prime.

The proof is by *reductio ad absurdum*, or reduction to a contradiction, which is one of the mathematician's favorite tools.

Once we know that the number of primes is infinite, we can ask ourselves whether there is any simple way of listing them in succession without missing a single one. A method of doing

FIGURE 9

this was first proposed by the ancient Greek philosopher and mathematician Eratosthenes and is usually known as "the sieve." All you have to do is to write the complete sequence of integers, 1, 2, 3, 4, etc., and then to strike out first all multiples of 2, then the remaining multiples of 3, then those of 5, etc. The sieve of Eratosthenes for the first hundred numbers is shown in Figure 9.

It contains altogether twenty-six primes. By using the above simple sieving method tables of primes up to one billion have been constructed.

It would be much simpler, however, if a formula could be devised by which we could find quickly and automatically only the primes and all the primes. But in spite of the attempts that have been made for centuries such a formula is still nonexistent. In 1640 the famous French mathematician Fermat thought that he had devised a formula that would produce only the prime numbers.

In his formula, $2^{2^n}+1$, n indicates the successive values of 1, 2, 3, 4, etc.

Using this formula we find:

$$2^2 +1=5$$
$$2^{2^2}+1=17$$
$$2^{2^3}+1=257$$
$$2^{2^4}+1=65537$$

Each of these is, in fact, a prime number. But about a century after Fermat's announcement the German mathematician Euler showed that in Fermat's fifth calculation, $2^{2^5}+1$, the result, 4,294,967,297, is not a prime, but is, in fact, the product of 6,700,417 and 641. Thus Fermat's empirical rule for calculating prime numbers proved to be wrong.

Another remarkable formula that produces many primes is:

$$n^2-n+41,$$

in which n again equals 1, 2, 3, etc. It has been shown that in all cases in which n indicates a number from 1 to 40 application of the above formula produces nothing but primes, but unfortunately it fails badly on the forty-first step.

In fact,

$$(41)^2-41+41=41^2=41\times41$$

which is a square, not a prime.

Still another attempted formula:

$$n^2-79n+1601$$

gives primes with n up to 79, but fails at 80!

Thus the problem of finding a general formula by the application of which only primes may be produced is still unsolved.

Another interesting example of a theorem of the theory of numbers that has been neither proved nor disproved is the so-called Goldbach conjecture, proposed in 1742, which states that *each even number can be represented as the sum of two primes.* You can easily find that it is true as applied to some simple examples, thus: $12 = 7 + 5$, $24 = 17 + 7$, and $32 = 29 + 3$. But, in spite of the immense amount of work done in this line, mathematicians have never been able either to give a conclusive proof of the infallibility of this statement or to find an example that would disprove it. As recently as 1931, a Russian mathematician, Schnirelman, succeeded in taking the first constructive step toward the desired proof. He was able to show that *each even number is the sum of not more than 300,000 primes.* Still more recently the gap between Schnirelman's "sum of three hundred thousand primes" and the desired "sum of two primes" was considerably narrowed by another Russian mathematician, Vinogradoff, who was able to reduce it to "the sum of four primes." But the last two steps from Vinogradoff's four to Goldbach's two primes seem to be the toughest of all, and nobody can tell whether another few years or another few centuries will be required to prove or disprove this difficult proposition.

Well, thus we seem still far away from deriving a formula that will give automatically all primes up to any desired large number, and there is even no assurance that such a formula ever will be derived.

We may now ask a more humble question—a question about the percentage of primes that can be found within a given numerical interval. Does this percentage remain approximately constant as we go to larger and larger numbers? And if not, does it increase or decrease? We can try to answer this question empirically by counting the number of primes as given in the tables. We find this way that there are 26 primes smaller than 100, 168 primes smaller than 1000, 78,498 primes smaller than 1,000,000, and 50,847,478 primes smaller than 1,000,000,000. Dividing these numbers of primes by corresponding numerical intervals we obtain the following table:

Interval 1–N	Number of primes	Ratio	$\dfrac{1}{\log._n n_N}$	Deviation %
1–100	26	0.260	0.217	20
1–1000	168	0.168	0.145	16
1–10^6	78498	0.078498	0.072382	8
1–10^9	50847478	0.050847478	0.048254942	5

This table shows first of all that the relative number of primes decreases gradually as the number of all integers increases, but that there is no point at which there are no primes.

Is there any simple way to represent mathematically this diminishing percentage of primes among large numbers? Yes there is, and the laws governing the average distribution of primes represents one of the most remarkable discoveries of the entire science of mathematics. It states simply *that the percentage of primes within an interval from 1 to any larger number N, is approximately stated by the natural logarithm of N.*[1] And the larger N is, the closer the approximation is.

In the table on this page you will find in the fourth column the natural logarithms of N. If you compare them with the values of the previous columns you will see that the agreement is fairly close and that the larger N is, the closer the agreement is.

As were many other propositions in the theory of numbers, the prime-number theorem given above was first discovered empirically and for a very long time was never confirmed by strict mathematical proof. It was not until nearly the end of the last century that the French mathematician Hadamard and the Belgian de la Vallée Poussin finally succeeded in proving it, by a method far too complicated and difficult to explain here.

This discussion of integers must not be dropped without mentioning the famous Great Theorem of Fermat, which will serve as an example of the class of problems not necessarily connected with the properties of prime numbers. The roots of this problem go back to ancient Egypt, where every good carpenter knew that a triangle with three sides in the ratio of 3:4:5 must include one right angle. In fact the ancient Egyptians used such a tri-

[1] In a simple way, a natural logarithm can be defined as the ordinary logarithm from the table, multiplied by the factor 2.3026.

angle, now known as an Egyptian triangle, as a carpenter's square.[2]

During the third century Diophantes of Alexandria began to wonder whether 3 and 4 were the only two integers the sum of whose squares would equal the squares of a third. He was able to show that there are other triplets (in fact an infinite number of them) of numbers having the same property, and gave a general rule for finding them. Such right-angled triangles in which all three sides are measured by integers are known now as Pythagorean triangles, the Egyptian triangle being the first of them. The problem of constructing Pythagorean triangles can be stated simply as an algebraic equation in which x, y, and z must be integers:[3]

$$x^2 + y^2 = z^2.$$

In the year 1621 Pierre Fermat in Paris bought a copy of the new French translation of Diophantes' book *Arithmetica*, in which Pythagorean triangles were discussed. When he read it, he made in the margin a short note to the effect that whereas the equation $x^2 + y^2 = z^2$ has an infinite number of integer solutions, *any equation of the type*

$$x^n + y^n = z^n,$$

where n is larger than 2, has no solution whatsoever.

"I have discovered a truly wonderful proof of this," added Fermat, "which, however, this margin is too narrow to hold."

When Fermat died, the book of Diophantes was discovered in his library and the contents of the marginal note became known

[2] The Pythagorean theorem of elementary school geometry states the proof thus:
$$3^2 + 4^2 = 5^2.$$

[3] Using the general rule of Diophantes (take any two numbers a and b such that $2ab$ is a perfect square. $x = a + \sqrt{2ab}$; $y = b + \sqrt{2ab}$; $z = a + b + \sqrt{2ab}$. Then $x^2 + y^2 = z^2$, which is easy to verify by ordinary algebra), we can construct the table of all possible solutions, the beginning of which runs:

$$3^2 + 4^2 = 5^2 \text{ (Egyptian triangle)}$$
$$5^2 + 12^2 = 13^2$$
$$6^2 + 8^2 = 10^2$$
$$7^2 + 24^2 = 25^2$$
$$8^2 + 15^2 = 17^2$$
$$9^2 + 12^2 = 15^2$$
$$9^2 + 40^2 = 41^2$$
$$10^2 + 24^2 = 26^2$$

to the world. That was three centuries ago, and ever since then the best mathematicians in each country have tried to reconstruct the proof that Fermat had in mind when he wrote his marginal note. But up to the present time no proof has been discovered. To be sure, considerable progress has been made toward the ultimate goal, and an entirely new branch of mathematics, the so-called "theory of ideals," has been created in attempts to prove Fermat's theorem. Euler demonstrated the impossibility of integer solution of the equations: $x^3 + y^3 = z^3$ and $x^4 + y^4 = z^4$, Dirichlet proved the same for the equation: $x^5 + y^5 = z^5$, and through the combined efforts of several mathematicians we now have proofs that no solution of the Fermat equation is possible when n has any value smaller than 269. Yet no general proof, good for *any* values of the exponent n, has ever been achieved, and there is a growing suspicion that Fermat himself either did not have any proof or made a mistake in it. The problem became especially popular when a prize of a hundred thousand German marks was offered for its solution, though of course all the efforts of money-seeking amateurs did not accomplish anything.

The possibility, of course, always remains that the theorem is wrong and that an example can be found in which the sum of two equal high powers of two integers is equal to the same power of a third integer. But since in looking for such an example one must now use only exponents larger than 269, the search is not an easy one.

2. THE MYSTERIOUS $\sqrt{-1}$

Let us now do a little advanced arithmetic. Two times two are four, three times three are nine, four times four are sixteen, and five times five are twenty-five. Therefore: the square root of four is two, the square root of nine is three, the square root of sixteen is four, and the square root of twenty-five is five.[4]

But what would be the square root of a negative number? Have expressions like $\sqrt{-5}$ and $\sqrt{-1}$ any meaning?

[4] It is also easy to find the square roots of many other numbers. Thus, for example, $\sqrt{5} = 2.236$ . . . . because: $(2.236$ . . . .$) \times (2.236$ . . . .$)$ $= 5.000$ . . . . and $\sqrt{7.3} = 2.702$ . . . . because: $(2.702$ . . . .$) \times (2.702$ . . . .$)$ $= 7.300$ . . . .

If you try to figure it out in a rational way, you will undoubtedly come to the conclusion that the above expressions make no sense at all. To quote the words of the twelfth century mathematician Brahmin Bhaskara: "The square of a positive number, as also that of a negative number, is positive. Hence the square root of a positive number is twofold, positive and negative. There is no square root of a negative number, for a negative number is not a square."

But mathematicians are obstinate people, and when something that seems to make no sense keeps popping up in their formulas, they will do their best to put sense into it. And the square roots of negative numbers certainly do keep popping up in all kinds of places, whether in the simple arithmetical questions that occupied mathematicians of the past, or in the twentieth century problem of unification of space and time in the frame of the theory of relativity.

The brave man who first put on paper a formula that included the apparently meaningless square root of a negative number was the sixteenth century Italian mathematician Cardan. In discussing the possibility of splitting the number 10 into two parts the product of which would be 40, he showed that, although this problem does not have any rational solution, one could get the answer in the form of two impossible mathematical expressions:

$5+\sqrt{-15}$ and $5-\sqrt{-15}$.[5]

Cardan wrote the above lines with the reservation that the thing is meaningless, fictitious, and imaginary, but still he wrote them.

And if one dares to write square roots of negatives, imaginary as they may be, the problem of splitting the number 10 into the two desired parts can be solved. Once the ice was broken the square roots of negative numbers, or imaginary numbers as they were called after one of Cardan's epithets, were used by various mathematicians more and more frequently, although always with great reservations and due excuses. In the book on algebra pub-

[5] The proof follows:
$(5+\sqrt{-15})+(5-\sqrt{-15})=5+5=10$ and
$(5+\sqrt{-15})\times(5-\sqrt{-15})=(5\times5)+5\sqrt{-15}-5\sqrt{-5}-(\sqrt{-15}\times\sqrt{-15})$
$=(5\times5)-(-15)=25+15=40.$

lished in 1770 by the famous German mathematician Leonard Euler we find a large number of applications of imaginary numbers, mitigated however, by the comment: "All such expressions as $\sqrt{-1}$, $\sqrt{-2}$, etc. are impossible or imaginary numbers, since they represent roots of negative quantities, and of such numbers we may truly assert that they are neither nothing, nor greater than nothing, nor less than nothing, which necessarily constitutes them imaginary or impossible."

But in spite of all these abuses and excuses imaginary numbers soon became as unavoidable in mathematics as fractions, or radicals, and one could practically not get anywhere without using them.

The family of imaginary numbers represents, so to speak, a fictitious mirror image of the ordinary or real numbers, and, exactly in the same way as one can produce all real numbers starting with the basic number 1, one can also build up all imaginary numbers from the basic imaginary unit $\sqrt{-1}$, which is usually denoted by the symbol i.

It is easy to see that $\sqrt{-9} = \sqrt{9} \times \sqrt{-1} = 3i$; $\sqrt{-7} = \sqrt{7} \cdot \sqrt{-1}$ $= 2.646 \ldots i$ etc., so that each ordinary real number has its imaginary double. One can also combine real and imaginary numbers to make single expressions such as $5 + \sqrt{-15} = 5 + i\sqrt{15}$ as it was first done by Cardan. Such hybrid forms are usually known as complex numbers.

For well over two centuries after imaginary numbers broke into the domain of mathematics they remained enveloped by a veil of mystery and incredibility until finally they were given a simple geometrical interpretation by two amateur mathematicians: a Norwegian surveyor by the name of Wessel and a Parisian bookkeeper, Robert Argand.

According to their interpretation a complex number, as for example $3 + 4i$, may be represented as in Figure 10, in which 3 corresponds to the horizontal distance, and 4 to the vertical, or ordinate.

Indeed all ordinary real numbers (positive or negative) may be represented as corresponding to the points on the horizontal axis, whereas all purely imaginary ones are represented by the points on the vertical axis. When we multiply a real number, say 3, representing a point on the horizontal axis, by the imaginary unit i we obtain the purely imaginary number $3i$, which must be plotted on the vertical axis. Hence, *the multiplication by i is geometrically equivalent to a counterclockwise rotation by a right angle.* (See Figure 10).

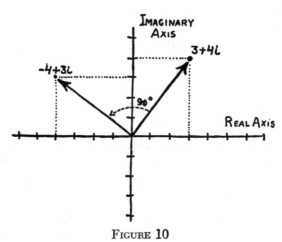

FIGURE 10

If now we multiply $3i$ once more by i, we must turn the thing by another 90 degrees, so that the resulting point is again brought back to the horizontal axis, but is now located on the negative side. Hence,

$$3i \times i = 3i^2 = -3, \text{ or } i^2 = -1.$$

Thus the statement that the "square of i is equal to -1" is a much more understandable statement than "turning twice by a right angle (both turns counterclockwise) you will face in the opposite direction."

The same rule also applies, of course, to hybrid complex numbers. Multiplying $3+4i$ by i we get:

$$(3+4i) i = 3i + 4i^2 = 3i - 4 = -4 + 3i.$$

And as you can see at once from Figure 10, the point $-4+3i$ corresponds to the point $3+4i$, which is turned counterclockwise by 90 degrees around the origin. Similarly the multiplication by

$-i$ is nothing but the clockwise rotation around the origin, as can be seen from Figure 10.

If you still feel a veil of mystery surrounding imaginary numbers you will probably be able to disperse it by working out a simple problem in which they have practical application.

There was a young and adventurous man who found among his great-grandfather's papers a piece of parchment that revealed the location of a hidden treasure. The instructions read:

"Sail to _____ North latitude and _____ West longitude[6] where thou wilt find a deserted island. There lieth a large meadow, not pent, on the north shore of the island where standeth a lonely oak and a lonely pine.[7] There thou wilt see also an old gallows on which we once were wont to hang traitors. Start thou from the gallows and walk to the oak counting thy steps. At the oak thou must turn *right* by a right angle and take the same number of steps. Put here a spike in the ground. Now must thou return to the gallows and walk to the pine counting thy steps. At the pine thou must turn *left* by a right angle and see that thou takest the same number of steps, and put another spike into the ground. Dig halfway between the spikes; the treasure is there."

The instructions were quite clear and explicit, so our young man chartered a ship and sailed to the South Seas. He found the island, the field, the oak and the pine, but to his great sorrow the gallows was gone. Too long a time had passed since the document had been written; rain and sun and wind had disintegrated the wood and returned it to the soil, leaving no trace even of the place where it once had stood.

Our adventurous young man fell into despair, then in an angry frenzy began to dig at random all over the field. But all his efforts were in vain; the island was too big! So he sailed back with empty hands. And the treasure is probably still there.

A sad story, but what is sadder still is the fact that the fellow might have had the treasure, if only he had known a bit about

[6] The actual figures of longitude and latitude were given in the document but are omitted in this text, in order not to give away the secret.

[7] The names of the trees are also changed for the same reason as above. Obviously there would be other varieties of trees on a tropical treasure island.

mathematics, and specifically the use of imaginary numbers. Let
us see if we can find the treasure for him, even though it is too
late to do him any good.

FIGURE 11

Treasure hunt with imaginary numbers.

Consider the island as a plane of complex numbers; draw one
axis (the real one) through the base of the two trees, and
another axis (the imaginary one) at right angles to the first,
through a point half way between the trees (Figure 11). Taking
one half of the distance between the trees as our unit of length,

we can say that the oak is located at the point -1 on the real axis, and the pine at the point $+1$. We do not know where the gallows was so let us denote its hypothetical location by the Greek letter Γ (capital gamma), which even looks like a gallows. Since the gallows was not necessarily on one of the two axes Γ must be considered as a complex number: $\Gamma = a + bi$, in which the meaning of a and b is explained by Figure 11.

Now let us do some simple calculations remembering the rules of imaginary multiplication as stated above. If the gallows is at Γ and the oak at -1, their separation in distance and direction may be denoted by $(-1) - \Gamma = -(1 + \Gamma)$. Similarly the separation of the gallows and the pine is $1 - \Gamma$. To turn these two distances by right angles clockwise (to the right) and counterclockwise (to the left) we must, according to the above rules multiply them by $-i$ and by i, thus finding the location at which we must place our two spikes as follows:

first spike: $(-i)[-(1+\Gamma)] + 1 = i(\Gamma + 1) - 1$
second spike: $(+i)(1-\Gamma) - 1 = i(1-\Gamma) + 1$

Since the treasure is halfway between the spikes, we must now find one half the sum of the two above complex numbers. We get:

$$\tfrac{1}{2}[i(\Gamma + 1) + 1 + i(1-\Gamma) - 1] = \tfrac{1}{2}[+i\,\Gamma + i + 1 + i - i\,\Gamma - 1]$$
$$= \tfrac{1}{2}(+2i) = +i.$$

We now see that the unknown position of the gallows denoted by Γ fell out of our calculations somewhere along the way, and that, regardless of where the gallows stood, the treasure must be located at the point $+i$.

And so, if our adventurous young man could have done this simple bit of mathematics, he would not have needed to dig up the entire island, but would have looked for the treasure at the point indicated by the cross in Figure 11, and there would have found the treasure.

If you still do not believe that it is absolutely unnecessary to know the position of the gallows in order to find the treasure, mark on a sheet of paper the positions of two trees, and try to carry out the instructions given in the message on the parchment by assuming several different positions for the gallows. You will always get the same point, corresponding to the number $+i$ on the complex plane!

Another hidden treasure that was found by using the imaginary square root of -1 was the astonishing discovery that our ordinary three-dimensional space and time can be united into one four-dimensional picture governed by the rules of four-dimensional geometry. But we shall come back to this discovery in one of the following chapters, in which we discuss the ideas of Albert Einstein and his theory of relativity.

PART II

Space, Time & Einstein

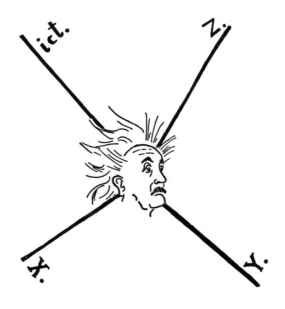

Unusual Properties of Space

1. *DIMENSIONS AND CO-ORDINATES*

WE ALL know what space is, although we should find ourselves in a rather awkward position if we were asked to define exactly what we mean by the word. We should probably say that space is *that* which surrounds us, and through which we can move forward or backward, right or left, up or down. The existence of the three independent mutually perpendicular directions represents one of the most fundamental properties of the physical space in which we live; we say that our space is three-directional or three-dimensional. Any location in space can be indicated by referring to these three directions. If we are visiting an unfamiliar city and we ask at the hotel desk how to find the office of a certain well-known firm, the clerk may say: "Walk five blocks south, two blocks to the right, and go up to the seventh floor." The three numbers just given are usually known as co-ordinates, and refer, in this case, to the relationship between the city streets, the building floors, and the point of origin in the hotel lobby. It is clear, however, that directions to the same location can be given from any other point, by using a co-ordinate system, which would correctly express the relationship between the new point of origin and the destination, and that the new co-ordinates can be expressed through the old ones by a simple mathematical procedure provided we know the relative position of the new co-ordinate system in respect to the old one. This process is known as *the transformation of co-ordinates*. It may be added here that it is not at all necessary that all three co-ordinates be expressed by the numbers representing certain distances; and, in fact, it is more convenient in certain cases to use angular co-ordinates.

Thus, for example, whereas addresses in New York City are most naturally expressed by a *rectangular* co-ordinate system

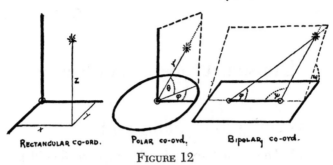

RECTANGULAR CO-ORD.   POLAR CO-ORD,   BIPOLAR CO-ORD.

FIGURE 12

represented by streets and avenues, the address system of Moscow (Russia) would certainly gain by being transformed into polar co-ordinates. This old city grew around the central fortress of the Kremlin, with radially diverging streets and several concentric circular boulevards, so that it would be natural to speak of a house located, say, twenty blocks north-north-west from the Kremlin wall.

Another classic example of a rectangular and a polar co-ordinate system is presented by the Navy Department building and by the War Department's Pentagon building in Washington, D. C., familiar to anybody connected with war work during World War II.

In Figure 12 we give several examples showing how the position of a point in space can be described in different ways by three co-ordinates some of which are distances and some, angles. But whatever system we choose we shall always need *three* data since we are dealing with a three-dimensional space.

Although it is difficult for us, with our three-dimensional concept of space, to imagine superspaces in which there are more than three dimensions (though, as we shall see later, such spaces exist), it is easy for us to conceive of a subspace, with fewer than three. A plane, a surface of a sphere, or as a matter of fact, any other surface is a two-dimensional subspace, since the position of a point on the surface can always be described by only two numbers. Similarly a line (straight or curved) is a one-dimensional subspace, and only one number is needed to describe a position on it. We can also say that a point is a subspace of zero dimensions, since there are no two different locations within a point. But who is interested in points anyway!

Being three-dimensional creatures we find it much easier to comprehend the geometrical properties of lines and surfaces, on which we can look "from the outside," than similar properties of three-dimensional space, of which we are ourselves a part. This explains why although you have no difficulty in understanding what is meant by a curved line, or a curved surface, you may yet be taken aback by the statement that three-dimensional space also can be curved.

However, with a little practice, and an understanding of what the word "curvature" really means, you will find the notion of a curved three-dimensional space very simple indeed, and toward the end of the next chapter, will (we hope!) be able even to speak with ease about what, at first sight, may seem a horrible notion, that is, a curved four-dimensional space.

But before we discuss that, let us try a few mental gymnastics with some facts about ordinary three-dimensional space, two-dimensional surfaces, and one-dimensional lines.

2. GEOMETRY WITHOUT MEASURE

Although your memory of the geometry with which you became familiar in your school days, that is, the science of space measurements,[1] may tell you that it consists mostly of a large number of theorems concerning the numerical relationships between various distances and angles (as, for example, the famous Pythagorean theorem concerning the three sides of a right-angled triangle), the fact is that a great many of the most fundamental properties of space do not require any measurements of lengths or angles whatsoever. The branch of geometry concerned with these matters is known as *analysis situs* or *topology*[2] and is one of the most provocative and difficult of the departments of mathematics.

To give a simple example of a typical topological problem, let

---

[1] The name geometry comes from two Greek words *ge*=earth, or rather ground, and *metrein*=to measure. Apparently, at the time the word was formulated, the ancient Greeks' interest in the subject was dominated by their real estate.

[2] Which means, from the Latin and the Greek respectively, the study of locations.

us consider a closed geometrical surface, say that of a sphere, divided by a network of lines into many separate regions. We can prepare such a figure by locating on the surface of a sphere an arbitrary number of points and connecting them with non-intersecting lines. What are the relationships that exist between the number of original points, the number of lines representing the boundaries between adjacent regions, and the number of regions themselves?

First of all, it is quite clear that if instead of the sphere we had taken a flattened spheroid like a pumpkin, or an elongated body like a cucumber, the number of points, lines, and regions would have been exactly the same on a perfect sphere. In fact, we

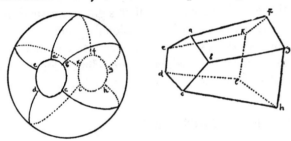

FIGURE 13

A subdivided sphere transformed into a polyhedron.

can take any closed surface that can be obtained by deforming a rubber balloon, by stretching it, by squeezing it, by doing to it anything we like, except cutting or tearing it, and neither the formulation nor the answer to our question will change in the slightest way. This fact presents a striking contrast to the facts of ordinary numerical relationships in geometry (such as the relationships that exist among linear dimensions, surface areas, and volumes of geometrical bodies). Indeed such relationships would be materially distorted if we stretched a cube into a parallelopiped, or squeezed a sphere into a pancake.

One of the things we can do with our sphere divided into a number of separate regions is to flatten each region so that the sphere becomes a polyhedron; the lines bounding different regions now become the edges of the polyhedron, and the original set of points become its vertices.

Our previous problem can now be reformulated, without however changing its sense, into a question concerning the relationships between the number of vertices, edges, and faces in a polyhedron of an arbitrary type.

In Figure 14 we show five regular polyhedrons, that is, those in which all faces have an equal number of sides and vertices, and one irregular one drawn simply from imagination.

In each of these geometrical bodies we can count the number of vertices, the number of edges, and the number of faces. What is the relation between these three numbers, if any?

By direct counting we can build the accompanying table.

Name	V number of vertices	E number of edges	F number of faces	V+F	E+2
Tetrahedron (pyramid)	4	6	4	8	8
Hexahedron (cube)	8	12	6	14	14
Octahedron	6	12	8	14	14
Icosahedron	12	30	20	32	32
Dodecahedron or Pentagon-dodecahedron	20	30	12	32	32
"Monstrosity"	21	45	26	47	47

At first the figures given in the three columns (under V, E, and F) do not seem to show any definite correlation, but after a little study you will find that the sum of the figures in the V and F columns always exceed the figure in the E column by two. Thus we can write the mathematical relationship:

$$V + F = E + 2.$$

Does this relationship hold for only the five particular polyhedrons shown in Figure 14, or is it also true for any polyhedron? If you try to draw several other polyhedrons different from those shown in Figure 14, and count their vertices, edges, and faces, you will find that the above relationship exists in every case. Apparently then, $V + F = E + 2$ is a general mathematical theorem of a topological nature since the relationship expression does not depend on measuring the lengths of the ribs, or the areas of the faces, but is concerned only with the number of the different geometrical units (that is, vertices, edges, faces) involved.

The relationship we have just found between the number of

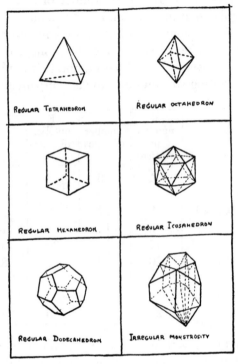

Regular Tetrahedron

Regular Octahedron

Regular Hexahedron

Regular Icosahedron

Regular Dodecahedron

Irregular Monstrosity

FIGURE 14

Five regular polyhedrons (the only possible ones) and one irregular monstrosity.

vertices, edges, and faces in a polyhedron was first noticed by the famous French mathematician of the seventeenth century, René Descartes, and its strict proof was demonstrated somewhat later by another mathematical genius, Leonard Euler, whose name it now carries.

Here is the complete proof of Euler's theorem, following the text of R. Courant and H. Robbins' book *What Is Mathematics?*,[3] just to show how things of that kind are done:

"To prove Euler's formula, let us imagine the given simple polyhedron to be hollow, with a surface made of thin rubber [Figure 15*a*]. Then if we cut out one of the faces of the hollow polyhedron, we can deform the remaining surface until it stretches [Figure 15*b*] out flat on a plane. Of course, the areas of the faces and the angles between the edges of the polyhedron will be changed in this process. But the network of vertices and edges in the plane will contain the same number of vertices and edges as did the original polyhedron, while the number of poly-

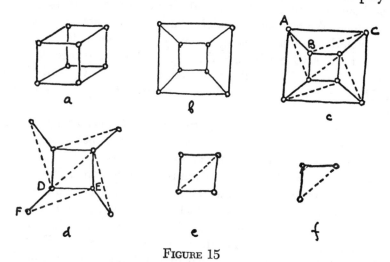

FIGURE 15

Proof of Euler's theorem. The drawing is made specifically for a cube, but the result would be the same if it were any other polyhedron.

gons will be one less than in the original polyhedron since one face was removed. We shall now show that for the plane network, $V-E+F=1$, so that, if the removed face is counted, the result is $V-E+F=2$ for the original polyhedron.

[3] The author is grateful to Drs. Courant and Robbins and to the Oxford University Press for permission to reproduce the passage that follows. Those readers who become interested in the problems of topology on the basis of the few examples given here will find a more detailed treatment of the subject in *What Is Mathematics?*

"First we 'triangulate' the plane network in the following way: In some polygon of the network which is not already a triangle we draw a diagonal. The effect of this is to increase both E and F by 1, thus preserving the value of $V-E+F$. We continue now drawing diagonals, joining pairs of points until the figure consists entirely of triangles, as it must eventually [Figure 15c]. In the triangulated network, $V-E+F$ has the same value as it had before the division into triangles, since the drawing of diagonals has not changed it.

"Some of the triangles have the edges on the boundary of the network. Of these, some, such as ABC, have only one edge on the boundary, while other triangles may have two edges on the boundary. We take any boundary triangle and remove that part of it which does not also belong to some other triangle [Figure 15d]. Thus from ABC we remove the edge AC and the face, leaving the vertices, A, B, C and the two edges AB and BC; while from DEF we remove the face, the two edges DF and FE, and the vertex F.

"The removal of a triangle of the type ABC decreases E and F by 1, while V is unaffected, so that $V-E+F$ remains the same. The removal of a triangle of type DEF decreases V by 1, E by 2 and F by 1, so that $V-E+F$ again remains the same. By a properly chosen sequence of these operations we can remove triangles with edges on the boundary (which changes with each removal), until finally only one triangle remains, with its three edges, three vertices, and one face. For this simple network, $V-E+F=3 -3+1=1$. But we have seen that by constantly erasing triangles $V-E+F$ was not altered. Therefore in the original plane network $V-E+F$ must equal 1 also, and thus equals one for the polyhedron with one face missing. We conclude that $V-E+F=2$ for the complete polyhedron. This completes the proof of Euler's formula."

One interesting consequence of Euler's formula is the proof that *there can be only five regular polyhedrons, namely those shown in Figure 14.*

In looking through the discussion of the last few pages carefully, you may notice, however, that in making the drawings of the polyhedrons "of all different kinds" shown in Figure 14, as well as in the mathematical reasoning leading to the proof of Euler's theorem, we made one hidden assumption that results in a considerable limitation of our choice. We have limited ourselves only to the polyhedrons that, so to speak, *do not have any holes through them;* and when we speak about holes, we do not mean something like a hole torn in a rubber balloon, but rather something like the hole in a doughnut or the enclosed hollow of a rubber tire tube.

A glance at Figure 16 will clarify the situation. We see here two different geometrical bodies, each of which is no less a polyhedron than any of the bodies shown in Figure 14.

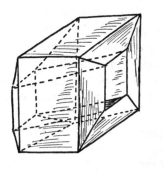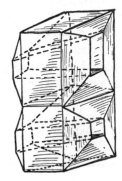

FIGURE 16

The two rivals of the ordinary cube with one and two holes through them. The faces are not all exactly rectangular but, as we have seen, this does not matter in topology.

Let us see now whether Euler's theorem is applicable to our new polyhedrons.

In the first case we count altogether 16 vertices, 32 edges, and 16 faces; thus $V+F=32$, whereas $E+2=34$. In the second case we have 28 vertices, 46 edges, and 30 faces so that $V+F=58$, whereas $E+2=48$. Wrong again!

Why is it so, and what is the reason that our general proof of Euler's theorem as given above fails in these cases?

The trouble is, of course, that whereas all the polyhedrons we have considered above can be related to a football bladder or balloon, the hollow polyhedrons of the new type are more like a

tire tube or still more complicated products of the rubber industry. To such polyhedrons as these latter the above given mathematical proof cannot be applied because with bodies of this kind
we cannot carry out all the operations necessary to the proof. In
fact, we have been asked: "to cut out one of the faces of the
hollow polyhedron, and to deform the remaining surface until it
stretches out flat on the plane."

If you take a football bladder and cut out with the scissors a
part of its surface you will have no trouble fulfilling that require-
ment. But you cannot do this successfully with a tire tube, no
matter how hard you try. If a glance at Figure 16 will not convince you of this, get an old tube and try!

You must not think, however, that there is no relationship
between the V, E, and F for the polyhedrons of the more complicated type; there is, but it is a different relationship. For the
doughnut-shaped, or, speaking more scientifically, torus-shaped,
polyhedrons we have $V+F=E$, whereas for the "pretzel" we
have $V+F=E-2$. In general $V+F=E+2-2N$ where N is the
number of holes.

Another typical topological problem closely connected with

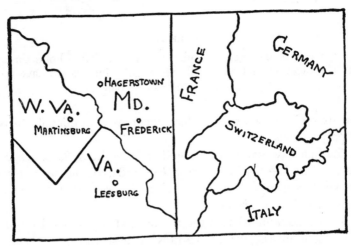

FIGURE 17

Topological maps of Maryland, Virginia, and West Virginia (on the
left) and Switzerland, France, Germany, and Italy (on the right).

Euler's theorem is the so-called "problem of the four colors." Suppose we have a surface of a sphere subdivided into a number of separate regions, and we are asked to color these regions in such a way that no two adjacent regions (that is, those having a common boundary) will have the same color. What is the smallest number of different colors we must use for such a task? It is clear that two colors only will in general not suffice since, when three boundaries come together in one point (as, for example, those of Virginia, West Virginia, and Maryland on a map of the United States, Figure 17) we shall need different colors for all of the three states.

It is also not difficult to find an example (Switzerland during the German annexation of Austria) where four colors are necessary (Figure 17).[4]

But try as you will, you never will be able to construct an imaginary map, be it on the globe or on a flat piece of paper,[5] for which more than four colors would be necessary. It seems that no matter how complicated we make the map, four colors always suffice to avoid any confusion along the boundaries.

Well, if this last statement is true one should be able to prove it mathematically, but in spite of the efforts of generations of mathematicians this has not yet been done. Here is a typical case of a mathematical statement that practically nobody doubts, but that nobody has been able to prove. The best that has been accomplished mathematically has been to prove that five colors are always sufficient. That proof is based on the Euler relationship, which has been applied to the number of countries, the number of their boundaries, and the number of triple, quadruple, etc. points in which several countries meet.

We do not demonstrate this proof, since it is fairly complicated and would lead us away from the main subject of the discussion, but the reader can find it in various books on topology and spend a pleasant evening (and perhaps a sleepless night) in contem-

[4] Before the annexation three colors would have sufficed: Switzerland, green; France and Austria, red; Germany and Italy, yellow.

[5] The cases of the plane map and that on the globe are the same from the point of view of the coloring problem, since, having the problem solved on a globe, we can always make a little hole in one of the colored regions and "open up" the resulting surface on the plane. Again a typical topological transformation.

plating it. Either he can try to devise the proof that not only five, but even four colors are sufficient to color any map, or, if he is skeptical about the validity of this statement, he can draw a map for which four colors are not enough. In the event of success in either of the two attempts his name will be perpetuated in the annals of pure mathematics for centuries to come.

Ironically enough, the coloring problem, which so successfully eludes solution for a globe or a plane, can be solved in a comparatively simple way for more complicated surfaces such as those of a doughnut or a pretzel. For example, it has been conclusively proved that seven different colors are enough to color any possible combination of subdivisions of a doughnut without ever coloring two adjacent sections the same, and examples have been given in which the seven colors are actually necessary.

In order to get another headache the reader may get an inflated tire tube and a set of seven different paints, and try to paint the surface of the tube in such a way that each region of a given color touches six other regions of different colors. After doing it, he will be able to say that "he really knows his way around the doughnut."

3. *TURNING SPACE INSIDE OUT*

So far we have been discussing the topological properties of various surfaces exclusively, that is, the subspaces of only two dimensions, but it is clear that similar questions can also be asked in relation to the three-dimensional space in which we ourselves live. Thus the three-dimensional generalization of the map-coloring problem can be formulated somewhat as follows: We are asked to build a space mosaic using many variously shaped pieces of different materials, and want to do it in such a way that no two pieces made of the same material will be in contact along the common surface. How many different materials are necessary?

What is the three-dimensional analogy of the coloring problem on the surface of a sphere or torus? Can one think about some unusual three-dimensional spaces that stand in the same relation to our ordinary space, as the surfaces of the sphere or torus to the ordinary plane surface? At first the question looks senseless. In fact, whereas we can easily think of many surfaces of various shapes, we are inclined to believe that there can be only one type of three-dimensional space, namely the familiar physical space in which we live. But such an opinion represents a dangerous delusion. If we stimulate our imaginations a little, we can think of three-dimensional spaces that are rather different from that studied in the textbooks of Euclidian geometry.

The difficulty in imagining such odd spaces lies mainly in the fact that, being ourselves three-dimensional creatures, we have to look on the space so to speak "from inside," and not "from outside" as we do with various odd-shaped surfaces. But with some mental gymnastics we will conquer these odd spaces without much trouble.

Let us first try to build a model of a three-dimensional space that would have properties similar to the surface of a sphere. The main property of a spherical surface is, of course, that, though it has no boundaries, it still has a finite area; it just turns around and closes on itself. Can we imagine a three-dimensional space that would close on itself in a similar way, and thus have a finite volume without having any sharp boundaries? Think about two spherical bodies each limited by spherical surfaces, as the body of an apple is limited by its skin.

Imagine now that these two spherical bodies are put "through one another" and joined along the outer surface. Of course we do not try to tell you that one can take two physical bodies, such as our two apples and squeeze them through each other so that their skins can be glued together. The apples would be squashed but would never penetrate each other.

One must rather think about an apple with an intricate system of channels eaten through it by worms. There must be *two* breeds of worm, say white and black ones, who do not like each other and never join their respective channels inside the apple although they may start them at adjacent points on the surface. An apple attacked by these two kinds of worm will finally look somewhat like Figure 18, with a double network of channels, tightly intertwined and filling up the entire interior of our apple.

But, although white and black channels pass very close to each other, the only way to get from one half of the labyrinth to the other is to go first through the surfaces. If you imagine the channels becoming thinner and thinner, and their number larger and larger, you will finally envisage the space inside the apple as being formed by the overlapping of two independent spaces connected only at their common surface.

FIGURE 18

If you do not like worms, you can think of a double system of enclosed corridors and stairways that could have been built, for example, inside the giant sphere at the last World's Fair in New York. Each system of stairways can be thought of as running through the entire volume of the sphere, but to get from some point of the first system to an adjacent point of the second system, one would have to go all the way to the surface of the sphere, where the two systems join, and then all the way back again. We say that two spheres overlap without interfering with each other, and a friend of yours could be very close to you in spite of the fact that in order to see him, and to shake his hand you would have to go a long way around! It is important to notice that the joining points of the two stairway systems would not actually differ from any other point within the sphere, since it would always be possible to deform the whole structure so that the joining points would be pulled inward and the points that were previously inside would come to the surface. The second important point about our model is that in spite of the fact that the total combined length of channels is finite, there are no "dead ends." You could move through the corridors and stairway on and on without being stopped by any wall or fence, and if you walked far enough you would inevitably find yourself at the point from which you started. Looking at the entire structure *from outside* one can say that a person moving through the labyrinth finally would come back to the point of his departure simply because the corridors gradually turned around, but for the people who were *inside*, and could not even know that such a thing as the "outside" existed, the space would appear as being of *finite size and yet without any marked boundaries*. As we shall see in one of the next chapters, this *"self-inclosed space of three dimensions"* that has no apparent boundaries and yet is not at all infinite was found very useful in the discussion of the properties of the universe at large. In fact, observations carried on at the very limit of telescopic power seem to indicate that at these giant distances space begins to curve, showing a pronounced tendency to come back and to close on itself in the same way as do the channels in our example of an apple eaten by the worms. But before we go on to these exciting problems, we have to learn a little more about other properties of space.

We are not yet quite through with the apple and the worms, and the next question we ask is whether it is possible to turn a worm-eaten apple into a doughnut. Oh no, we do not mean to make it taste like a doughnut, but just to make it look like one. We are discussing geometry, and not the art of cooking. Let us take a double apple such as that discussed in the previous section, that is, two fresh apples put "through one another" and "glued together" along their surfaces. Suppose a worm has eaten within one of the apples a broad circular channel as shown in Figure 19. Within *one* of the apples, mind you, so that whereas outside the channel each point is a double one belonging to both apples, inside the channel we have only the material of the apple not eaten by the worm. Now our "double apple" has a free surface composed of the inner walls of the channel (Figure 19a).

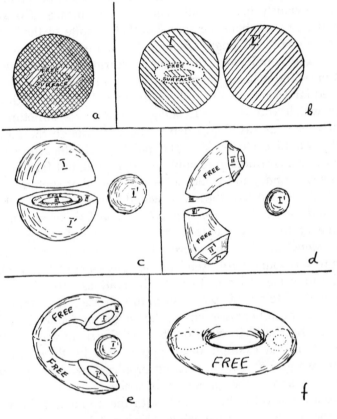

FIGURE 19

How to turn a double apple eaten by a worm into a good doughnut.
No magic; just topology!

Can you change the form of this spoiled apple so as to turn it into a doughnut? It is assumed, of course, that the material of the apple is quite plastic so that you can mold it any way you like, the only condition being that no rupture of the material must take place. To facilitate the operation, we may cut the material of the apple, provided we glue it back again after the required deformation is completed.

We start the operation by unfastening the skins of two parts forming the "double apple" and taking them apart (Figure 19*b*). We shall mark the two unglued surfaces by numerals I and I', in order to keep track of them in the following operations, so that we may glue them back in place again before we are finished. Now, cut the part containing the worm-eaten channel across so that the cut will go across the channel (Figure 19*c*). This operation opens two newly cut surfaces which we mark by II, II' and III, III', so that we shall know exactly where to fasten them together later. It also brings out the free surfaces of the channel, which is destined to form the free surface of the doughnut. Now, take the cut parts and stretch them in the way shown in Figure 19*d*. The free surface is now stretched out to a large extent (but according to our assumption the materials used are perfectly stretchable!). At the same time the cut surfaces I, II, and III have been reduced in their dimensions. While we are operating on the first half of the "double apple," we must also reduce the size of the second half squeezing it down to the dimensions of a cherry. Now, we are ready to start gluing back along the cuts we made. First, and that is easy, join the surfaces III, III' again, thus obtaining the shape shown in Figure 19*e*. Next, put the shrunken half apple between the two ends of the pincer thus formed, and bring the ends together. The surface of the ball marked I' will be glued up to the surfaces I from which it was originally unglued, and the cut surfaces II and II' will close on each other. As a result we get a doughnut, nice and smooth.

What's the point of all this?

None whatever, save to give you an exercise in imaginative geometry, a form of mental gymnastics that will help you understand such unusual things as curved space and space closed on itself.

If you want to stretch your imagination a bit farther, here is a "practical application" of the above procedure.

Your body also has the shape of a doughnut, though you probably never thought about it. In fact, in the very early stage of its development (embryonic stage) every living organism passes the stage known as "gastrula," in which it possesses a spherical shape with a broad channel going across it. Through one end of the channel food is taken in, through the other what is left of it after the body has used what it can, goes out. In fully developed organisms the internal channel becomes much thinner and more complicated, but the principle remains the same: and all geometrical properties of a doughnut remain unchanged.

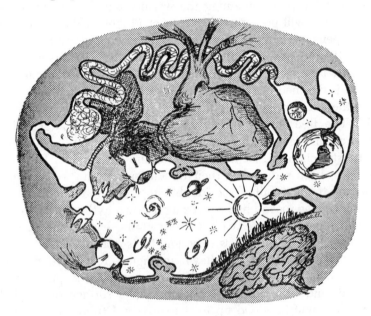

FIGURE 20

Inside-out universe. This surrealistic drawing represents a man walking on the surface of the Earth and looking up at the stars. The picture is transformed topologically according to the method indicated in Figure 19. Thus the Earth, sun, and stars are crowded in a comparatively narrow channel running through the body of the man, and surrounded by his internal organs.

Well, since you are a doughnut, try to make a transformation the reverse of that shown in Figure 19—try to transform your body (mentally!) into a double apple with a channel within. In particular, you find that whereas different parts of your body, partially overlapping one another, will form the body of the "double apple," the entire universe, including the earth, moon, sun, and stars, will be squeezed into the inner circular channel!

Try to draw a picture of how it will look, and if you do it well Salvador Dali himself will recognize your superiority in the art of surrealistic painting! (Figure 20).

We cannot conclude this section, long as it is, without some discussion of right- and left-handed bodies and their relation to the general properties of space. The problem may be introduced in the most convenient way by referring to a pair of gloves. If you compare two gloves of a pair (Figure 21) you will find them

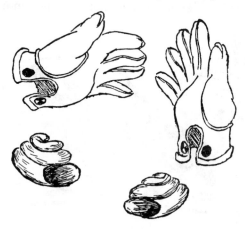

FIGURE 21

The right- and left-hand objects seem exactly alike but yet are quite different.

identical in all measurements and yet there is a great difference since you cannot put the left glove on the right hand or vice versa. You can turn and twist them as much as you like, but still the right glove remains right, and the left glove remains left. The same difference between right- and left-handed objects can be noticed in the construction of shoes, the steering mechanism of automobiles (American and British varieties), golf clubs, and many other things.

On the other hand, such things as men's hats, tennis rackets, and many other objects do not show such differences; nobody would be silly enough to order from a shop a dozen left-handed teacups, and it is certainly monkey business if someone asks you to borrow a left-handed monkey wrench from a neighbor. What is the difference between these two kinds of objects? If you think about it a little you will notice that objects like hats or teacups possess what we call a plane of symmetry along which they can be cut into two identical halves. No such plane of symmetry exists for gloves or shoes, and try as you will you will not be able to cut a glove into two identical parts. If the object does not possess a plane of symmetry, and is as we say, asymmetrical, it will be bound in two different modifications—a right- and a left-handed one. This difference occurs not only in man-made objects

FIGURE 22

An idea of two-dimensional "shadow-creatures" living on a plane. This kind of two-dimensional creature is not very "practical." The man has his face and not his profile, and cannot put into his mouth the grapes he holds in his hand. The donkey can eat the grapes all right but can walk only to the right and has to back in order to move to the left. It isn't unusual for donkeys, but not good in general.

like gloves or golf clubs, but also very often in nature. For example, there are two varieties of snails, which are identical in all other respects, but differ in the way they build their house: one variety has the shell spiraling clockwise, whereas the other spirals in a counterclockwise way. Even the so-called molecules, the tiny particles from which all different substances are built, often possess right- and left-handed forms, very similar to those of right and left gloves, or clockwise and counterclockwise snail shells.

You cannot see the molecules, of course, but the asymmetry shows up on the form of the crystals, and some optical properties of these substances. There are, for example, two different kinds of sugar, a right- and a left-handed sugar, and, believe it or not, there are also two kinds of sugar-eating bacteria, each kind consuming only the corresponding kind of sugar.

As was said above, it seems quite impossible to turn a right-handed object, a glove for example, into a left-handed one. But is that really true? Or can one imagine some tricky kind of space in which this can be done? To answer this question, let us examine it from the point of view of the flat inhabitants of a surface that can be observed by us from our superior three dimensional outlook. Look at Figure 22, representing some examples of the possible inhabitants of flatland, that is, of the space of two dimensions only. The man standing with a bunch of grapes in his hand can be called a "face-man" since he has a "face" but no "profile." The animal is, however, a "profile-donkey" or to be more specific a "right-looking-profile-donkey." Of course we can also draw a "left-looking-profile-donkey" and, since both donkeys are confined to the surface, they are just as different, from the two-dimensional point of view, as a right and a left glove in our ordinary space. You cannot superimpose a "left donkey" on a "right donkey," since in order to bring their noses and tails together you would have to turn one of them upside down, and thus his legs would be hitting the air instead of standing firmly on the ground.

But if you take one donkey out of the surface, turn it around in space, and put it back again, the two donkeys will become identical. By way of analogy one could say that a right glove can be turned into a left glove by taking it out of our space in the fourth direction and rotating it in a proper way before putting it back. But our physical space hasn't a fourth dimension, and the above described method must be considered as quite impossible. Isn't there any other way?

Well, let us return again to our two-dimensional world, but, instead of considering an ordinary plane surface as in Figure 22, investigate the properties of the so-called "surface of Möbius." This surface, named for a German mathematician who studied it first almost a century ago, can be easily made by taking a long strip of ordinary paper and gluing it into a ring, twisting it once before the two ends are joined together. Examination of Figure 23 will show you how to do it. This surface has many peculiar properties, one of which can be easily discovered by cutting, with a pair of scissors, completely around it in a line parallel to the edges (along the arrows in Figure 23). You would expect, of course, that by doing so, you would cut the ring into two separate rings. Do it, and you will see that your guess was wrong: instead of two rings you will find only one ring, but one twice as long as the original and half as wide!

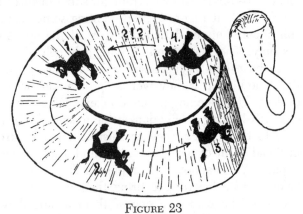

FIGURE 23
Surface of Möbius and Klein's bottle.

Let us see now what happens to a shadow donkey when he walks around on the Möbius surface. Suppose he starts with the position 1 (Figure 23) being seen at this moment as a "left-profile donkey." On and on he goes, passing through the positions 2 and 3, clearly visible in the picture, and finally approaches the spot from which he started. But to your, and his, surprise, our donkey finds itself (position 4) in an awkward position, his legs sticking up into the air. He can, of course, turn in his surface, so that his legs will come down, but then he will be facing the wrong way.

In short, by walking around the surface of Möbius, our "left-profile" donkey has turned into one with a "right profile." And, mind you, this has happened in spite of the fact that the donkey has remained on the surface all the time and hasn't been taken up and turned around in space. Thus we find that *on a twisted surface a right-hand object can be turned into a left-hand one, and vice versa, by merely carrying it around the twist.* The Möbius strip shown in Figure 23 represents a part of a more general surface, known as the Klein bottle (shown on the right in Figure 23), which has only one side and closes itself, having no sharp boundaries. If this is possible on a two-dimensional surface, the same must be true also in our three-dimensional space provided of course that it is twisted in a proper way. Naturally it is not easy to imagine a Möbius twist in space. We cannot look at our space from outside, as we looked at the donkey's surface, and it is always difficult to see things clearly when you are right in the midst of them. But it isn't at all impossible that astronomical space is closed on itself and in addition twisted in the Möbius way.

If this is really so travelers around the universe would come back left-handed with their hearts in the right part of their chests, and the manufacturers of gloves and shoes would have the dubious advantage of being able to simplify the production by making only one kind of shoes and gloves, and shipping one half of them around the universe to turn them into the kind needed for the other half of the world's feet.

On this fantastic thought, we finish our discussion of the unusual properties of unusual spaces.

The World of Four Dimensions

1. *TIME IS A FOURTH DIMENSION*

THE concept of the fourth dimension is usually surrounded by mystery and suspicion. How dare we, creatures of length, height, and width, speak of four-dimensional space? Is it possible by using all our three-dimensional intelligence to imagine a super-

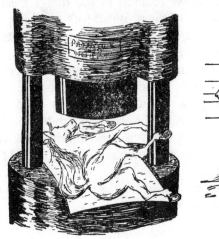

FIGURE 24

A wrong and a correct way to "squeeze" a three-dimensional body into a two-dimensional surface.

space of four dimensions? And what would a four-dimensional cube or sphere look like? When we say "imagine" a giant dragon with a long scaled tail and flame streaming from his nostrils, or a super-airliner with a swimming pool and a couple of tennis courts on its wings, you are actually drawing a mental picture of the way it would look were it to appear suddenly in front of you.

And you draw that picture against the background of the familiar three-dimensional space in which all ordinary objects, including yourself, are located. If this is the meaning of the word "imagine," then it is just as impossible to imagine a four-dimensional figure against the background of ordinary three-dimensional space, as

FIGURE 25

Two-dimensional creatures looking with surprise at the shadow of a three-dimensional cube projected onto their surface.

it is impossible to squeeze a three-dimensional body into a plane. But wait a moment. We *do*, in a certain sense, squeeze three-dimensional bodies into a plane by drawing a picture of them. In all these cases, however, we do not of course use a hydraulic press or any other physical force to do the job, but apply the method known as geometrical "projection" or shadow building.

The difference between the two ways of squeezing a body (for instance, that of a horse) into a plane can be at once understood by looking at Figure 24.

By way of analogy we can now say that although it is not possible to "squeeze" a four-dimensional body into a three-dimensional space without some parts sticking out, one can speak of the "projections" of various four-dimensional figures in our space of only three dimensions. But one must remember that just as the plane projections of three-dimensional bodies are two-dimensional or plane figures, so the projections of four-dimensional superbodies in our ordinary space will be represented by space-figures.

To make the matter clearer, let us first think how the two-dimensional shadow creatures living on a surface would conceive the idea of a three-dimensional cube; we can easily imagine that, since, being superior three-dimensional beings, we can look from above, that is, from the third direction, on the world of two dimensions. The only way to "squeeze" a cube into a plane is to "project" it on that plane in the way shown in Figure 25. Watching such a projection, and various other projections that can be obtained by rotation of the original cube, our two-dimensional friends will be able at least to form some idea about the properties of the mysterious figure called "a three-dimensional cube." They will not be able to "jump out" of their surface and visualize the cube the way we do, but by merely watching the projection they would be able to say, for example, that the cube has eight vertices and twelve edges. Now look at Figure 26, and you will find yourself exactly in the same situation as the poor two-dimensional shadow creatures inspecting the projection of an ordinary cube on their surface. In fact the strangely complicated structure that is being examined with such surprise by the members of the family, is actually the projection of a four-dimensional supercube in our ordinary three-dimensional space.[1]

Examine this figure carefully and you will easily recognize the same features as those puzzling the shadow creatures in Figure 25: whereas the projection of an ordinary cube on a plane is

---

[1] To be more exact, Figure 26 gives the projection on the plane of the paper of the projection in our space of a four-dimensional supercube.

represented by two squares, one inside the other, connected vertex to vertex, the projection of a supercube in ordinary space is formed by two cubes, one placed inside the other with their vertices connected in a similar way. And by counting you can easily see that a supercube has altogether 16 vertices, 32 edges and 24 faces. Quite a cube, is it not?

Now let us see what a four-dimensional sphere looks like. To do that we had better turn again to a more familiar case, that of a projection of an ordinary sphere on a plane surface. Think for example of a transparent globe, with the continents and oceans marked on it, being projected on a white wall (Figure 27). In

Figure 26

A visitor from the Fourth Dimension! A straight projection of a four-dimensional supercube.

the projection the two hemispheres will of course overlap each other, and, judging from the projection, one might think that the distance from New York (U. S. A.) to Peiping (China) is very short. But that is only an impression. In fact every point on the projection represents actually two opposite points on the actual sphere, and a projection of an airliner flying from New York to China on the globe, will move all the way to the rim of the plane projection, and then all the way back again. And in spite of the fact that the projections of two different airliners may overlap on the picture, no collision will take place if the airliners are "actually" on the opposite sides of the globe.

Such are the properties of the plane projection of an ordinary

sphere. Straining our imagination a little we shall have no difficulty in seeing how the space projection of a four-dimensional
supersphere looks. Just as the plane projection of an ordinary
sphere is formed by two flat discs put together (point to point)
and joined only along the outer circumference, the space-projection of a supersphere must be imagined as two spherical bodies
put through each other and joined along their outer surfaces.
But we have already discussed an extraordinary structure such
as this in the previous chapter, as an example of a closed three-
dimensional space analogous to a closed spherical surface. Thus
all we have to add here is that the three-dimensional projection

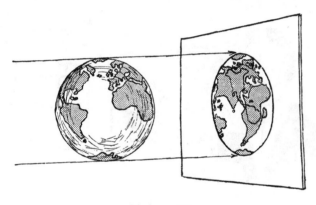

FIGURE 27

Plane projection of the globe.

of a four-dimensional sphere is nothing more than the Siamese-
twin-like apples we discussed there, formed by two ordinary
apples grown together along their entire skin surfaces.

In a similar way, by using the method of analogy, we can
answer many other questions concerning the properties of four-
dimensional figures, although try as we may, we will never be
able to "imagine" a fourth independent direction in our physical
space.

But if you think a little more about it, you will find that it is
not at all necessary to become mystical in order to conceive a
fourth direction. Indeed there is a word that most of us use
every day to designate that which could, and actually should, be

considered as a fourth independent direction in the physical world. We are talking here of time, which, together with space, is constantly used to describe events taking place around us. When we talk about any kind of happening in the universe, whether it is a casual encounter with a friend on the street or the explosion of a distant star, we usually not only say where it took place, but also when. Thus we add one more fact, the date, to the three directional facts that enter into our location of place.

If you consider the matter further you will also easily realize that each physical object has four dimensions, three in space and one in time. Thus the house in which you live extends so much in length, width, height, and time, the last extension being measured by the period of time from the date the house was built to the date it will finally burn down, or be taken apart by some wrecking company, or disintegrate at the end of an advanced old age.

To be sure, the direction of time is not quite the same as the three directions in space. Time intervals are measured by the clock, which makes ticktock sounds to denote seconds and ding-dong sounds to denote hours, in contrast to space intervals, which are measured by yardsticks. And whereas you can use the same yardstick to measure length, width, and height, you cannot turn a yardstick into a clock to measure duration of time. Also, whereas you can move forward, or to the right, or upward in space, and then come back again, you cannot come back in time, which drives you forcibly from the past into the future. But granting all these differences between the time-direction and the three directions in space, we can still use time as the fourth direction in the world of physical events, being careful however not to forget that it is not quite the same.

In choosing time as the fourth dimension we shall find it much simpler to visualize the four-dimensional figures discussed in the beginning of this chapter. Remember, for example, the strange figure cut by the projection of a four-dimensional cube? Sixteen vertices, thirty-two ribs, and twenty-four sides! No wonder that the people in Figure 26 are staring with such surprise at this geometrical monster.

From our new point of view, however, a four-dimensional cube is an ordinary cube that exists for a certain period of time. Suppose that you built a cube from twelve pieces of straight wire on the first of May, and took it apart one month later. Each corner point of such a cube must now be considered as being actually a line extending in the direction of time for the length of one month. You can attach a little calendar to each vertex and turn over the leaves each day to show the progress in time.

Now it is easy to count the number of ribs in our four-dimensional figure. You have, in fact, twelve space-ribs at the beginning of its existence, eight "time-ribs" representing the duration

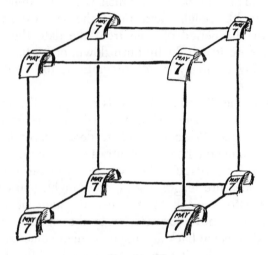

FIGURE 28

of each vertex, and again twelve space ribs at the end of the existence.[2] Altogether thirty-two ribs. In a similar way, we count that there are altogether sixteen vertices: eight space vertices on May 7 and again the same eight space-vertices on June 7. We leave it as an exercise for the reader to count the number of faces on our four-dimensional figure in the same way. In doing so it must be remembered that some of these faces will be ordinary square faces of the original cube, whereas the others will be

[2] If you do not understand this think of a square with four corner points, and four sides, which we move a certain distance perpendicularly to its surface (in the third direction) by a distance equal to its sides.

"half-space-half-time" faces formed by the original ribs of our cube extending in time from May 7 to June 7.

What we have said here about a four-dimensional cube can, of course, be applied to any other geometrical figure, or to any material object dead or alive.

In particular, think of yourself as a four-dimensional figure, a kind of long rubber bar extending in time from the moment of your birth to the end of your natural life. Unfortunately one cannot draw four-dimensional things on paper, so that in Figure 29 we have tried to convey this idea by an example of the two-dimensional shadow man taking for the time-direction the space-direction perpendicular to the two-dimensional plane on which he lives. The picture represents just a small section of the entire life span of our shadow man. The entire life span should be

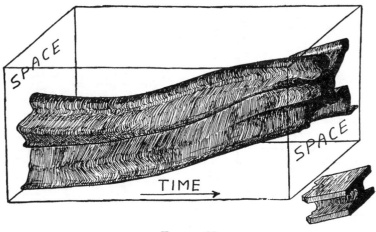

FIGURE 29

represented by a much longer rubber bar, which is rather thin in the beginning, when the man is still a baby, runs wiggling through the period of many years of life, attains a constant shape at the moment of death (because the dead do not move), and then begins to disintegrate.

To be more exact we must say that this four-dimensional bar is formed by a very numerous group of separate fibers, each one composed of separate atoms. Through the period of life most of these fibers stay together as a group; only a few of them fall away, as when the hair or the nails are cut. Since the atoms are indestructible, the distintegration of the human body after death should be actually considered as the dispersion of the separate filaments (except probably those forming the bones) in all different directions.

In the language of four-dimensional space-time geometry the line representing the history of each individual material particle

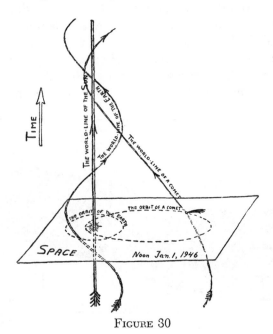

FIGURE 30

is known as its "world-line." Similarly we can speak of the "world-bands" composed of a group of world-lines forming a composite body.

In Figure 30 we give an astronomical example showing the world-lines of the sun, the earth, and a comet.[3] Here as in the previous example of a jumping man, we took two-dimensional space (the plane of the earth's orbit) and directed the time-axis perpendicular to it. The world-line of the sun is represented in

[3] Properly speaking we should speak here of "world-bands," but from the astronomical point of view one may consider stars and planets as points.

this graph by a straight line parallel to the time-axis, since we consider the sun as not moving.[4] The world-line of the earth, which moves on a very closely circular orbit, is a spiral winding around the sun-line, whereas the world-line of a comet approaches the sun-line and then goes far away again.

We see that from the point of view of four-dimensional space-time geometry the topography and the history of the universe fuse into one harmonious picture, and all we have to consider is a tangled bunch of world lines representing the motion of individual atoms, animals, or stars.

2. TIME-SPACE EQUIVALENT

In considering time as the fourth dimension more or less equivalent to the three spatial dimensions we run into one rather difficult question. When we measure length, width, or height we can use in all three cases one and the same unit, say 1 in. or 1 ft. But time-duration cannot be measured either in feet or in inches, and we have to use entirely different units, say minutes or hours. How do they compare? If we envisage a four-dimensional cube that measures in space 1 ft by 1 ft by 1 ft, how long must it extend in time to make all of our four dimensions equal? One sec, 1 hr, or 1 month as we have assumed in our previous example? Is 1 hr longer than 1 ft or is it shorter?

At first the question sounds meaningless, but if you think about it a little more you find a reasonable way in which a length and a duration may be compared. You often hear it said that someone lives "within twenty minutes of downtown by bus" or that some place is "only five hours away by train." Here we specify distances by giving the time necessary to cover them using a given type of transportation.

Thus, if we could agree on some *standard velocity* we should be able to express time intervals in units of length, or vice versa. It is clear, of course, that the standard velocity to be chosen as the fundamental translation factor between space and time must

---

[4] Actually our sun is moving in respect to the stars so that in reference to the stellar system the world-line of the sun should be somewhat inclined to one side.

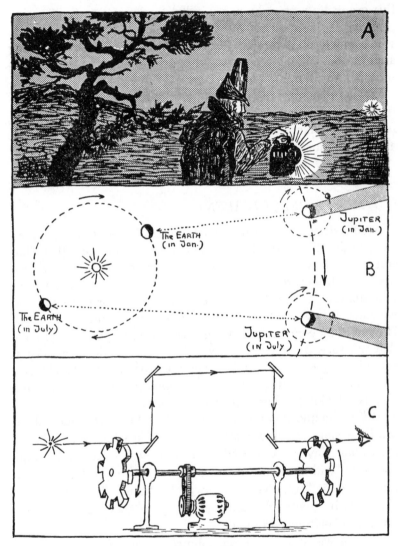

FIGURE 31

be of equally fundamental and general nature, being always the same regardless of human initiative or physical circumstances. The only velocity known in physics to possess the desired degree of generality is the velocity of light spreading through empty space. Though usually known as the "velocity of light" it can be better described as the "propagation velocity of physical interactions," since any kind of forces that act between material bodies, whether the forces of electric attraction or the forces of gravity, spread through empty space with the same speed. Besides, as we shall see later, the velocity of light represents the upper limit of any possible material velocity, and no object can travel through the space with a velocity greater than that.

The first attempt to measure the velocity of light was made by the famous Italian scientist Galileo Galilei in the seventeenth century. On a dark night Galileo went with his assistant into the open fields near Florence, taking along two lanterns each equipped with a mechanical shutter. The two men took positions a few miles apart from each other, and at a certain moment Galileo opened his lantern flashing a beam of light in the direction of his assistant (Figure 31A). The latter had been instructed to open his lantern as soon as he saw the light signal coming from Galileo. Since the light must have taken some time to come from Galileo to the assistant, and back to Galileo, it was expected that there would be a certain delay between the moment Galileo opened his lantern and the moment he saw the response coming back from his assistant. A small delay was actually noticed, but when Galileo sent his assistant to a position twice as far away, and repeated the experiment no increase of the delay was observed. Apparently the light traveled so rapidly that it took practically no time to cover the distance of a few miles, and the observed delay was caused by the fact that Galileo's assistant could not open his lantern at exactly the same moment as he saw the light—the delay of reflexes we call it now.

Although Galileo's experiment did not lead to any positive result, one of his other discoveries, namely that of the moons of Jupiter, supplied the basis for the first actual measurement of the speed of light. In the year 1675 the Danish astronomer Roemer, observing the eclipses of Jupiter's moons, noticed that the time intervals between the moments when the moons disappear in the shadow thrown by the planet are not always the same but appear shorter or longer depending on the distance between Jupiter and the earth at that particular time. Roemer realized immediately

(as you will after inspecting Figure 31B) that this effect is not
produced by any irregularity in the motion of Jupiter's moons,
but is simply due to the fact that we see these eclipses with
different delays because of the variable distance between Jupiter
and the earth. From his observations we were able to find that
the speed of light is about one hundred and eighty-five thousand
miles per second. No wonder that Galileo could not measure the
speed of light by his device since the light from his lantern
needed only a few hundred thousandths of a second to travel to
his assistant and back!

But what Galileo could not do with his rudimentary shutter
lanterns was done later by using more refined physical instruments. In Figure 31C we see the arrangement first used by the
French physicist Fizeau for measuring the speed of light at comparatively small distances. The main part of his arrangement
consists of two cogwheels set on a common axis in such a way
that if you look at the wheels parallel to the axis you can see the
cogs of the first wheel covering the intervals between the cogs
of the second one. Thus a thin beam of light sent parallel to the
axis cannot pass through, no matter how the axis is turned. Suppose now that the system of these two cogwheels is set into a
rapid rotation. Since the light passing between two cogs of the first
wheel must take some time before it reaches the second wheel,
it will be able to pass through if during that time the cogwheel
system turned by half the distance between two cogs. The situation here is rather similar to that of a car moving at a proper
speed along an avenue with a synchronized system of stop-lights.
If the wheels are rotating twice as fast, the second cog will come
into place by the time the light gets there, and its progress will
be again stopped. But at a still higher rotation speed the light
will be able to go through again since the cog will have passed
the path of the light, and the following opening will be within
the path of light just at the proper time to let it through. Thus,
noticing the rotation speeds corresponding to successive appear-
ances and disappearances of light one is able to estimate the
speed of light while traveling between the two wheels. To help
the experiment, and to reduce the necessary speed of rotation,
one can force the light to cover a larger distance while going

from the first cogwheel to the second; this can be done with the help of mirrors as indicated in Fig. 31C. In this experiment Fizeau found that he was first able to see light through the openings in the wheel nearest him when the apparatus was rotating at 1000 revolutions per second. This proved that at that speed cogs had traveled half the distance between them in the length of time necessary for light to travel the distance from one wheel to the other. Since each wheel had 50 cogs all of identical size, this distance was obviously 1/100 the circumference of the wheel, and the time of travel the same fraction of the time that it took the wheel to make a complete revolution. Relating these calculations to the distance through which the light passed from one wheel to the other, Fizeau arrived at a speed of 300,000 km, or 186,000 miles a second, which is about the same as the result obtained by Roemer in his observation of the satellites of Jupiter.

Following the work of these pioneers, a great number of independent measurements have been made using the methods of both astronomy and physics. The best estimate available at present of the speed of light through space (usually denoted by the letter "c") is

$$c = 299,776 \ \frac{\text{km}}{\text{sec}} \ \text{or} \ 186,300 \ \frac{\text{miles}}{\text{sec}}.$$

This tremendously high velocity of light makes it a convenient standard by which to measure astronomical distances so vast that to express them in miles or kilometers would be to deal with numerical notations that would fill whole pages. Thus, the astronomer will say that a certain star is 5 "light-years" away in the same sense that we speak of a place that is 5 hours away by train. Since a year contains 31,558,000 sec, one light-year corresponds to $31,558,000 \times 299,776 = 9,460,000,000,000$ km, or 5,879,000,000,000 miles. In this use of the term "light-years" to denote a measurement of distance we have a practical recognition of time as a dimension, and time units as a measurement of space. We can also reverse the procedure and speak of "light-miles," meaning the time necessary for light to cover the distance of one mile. Using the above value of light velocity, we find that one light-mile is equal to 0.0000054 sec. Similarly one "light-foot" is 0.0000000011 sec. This answers our question about the four-dimensional cube discussed in the previous section. If the space-dimensions of this cube are 1 ft by 1 ft by 1 ft, its space-duration must be only about 0.000000001 sec. If the space-cubic-foot exists for an entire month, it must be rather considered as the four-dimensional bar strongly elongated in the direction of the time-axis.

3. FOUR-DIMENSIONAL DISTANCE

Having settled the question concerning comparable units to be used along the space- and the time-axis, we can now ask ourselves what should be understood by the distance between two points in the four-dimensional space-time world. It must be remembered that each point in this case corresponds to what is usually known as "*an event*," that is the combination of the position and the time-date. To clarify the matter let us consider for example the following two events:

Event I. A bank located on the first floor at the corner of Fifth Avenue and 50th Street in New York City was robbed at 9:21 A.M. July 28, 1945.[5]

Event II. An army plane lost in the fog crashed into the 79th floor wall of the Empire State Building at 34th Street between Fifth and Sixth Avenues, New York City, at 9:36 A.M. the same day (Figure 32).

These two events were separated in space by 16 north-and-south blocks, 1/2 an east-and-west block, and 78 floors, and in time by 15 min. Obviously it is not necessary, in order to describe the space-separation between the two events, to note the individual numbers of avenue-blocks and of floors, since we can combine them into a single straight distance by means of the well-known Pythagorean theorem, according to which the distance between two points in space is the square root of the sum of the squares of the individual co-ordinate distances (Figure 32, corner). In order to apply the Pythagorean theorem, we must, of course, first express in comparable units, such as feet, all distances involved. If the length of a north-and-south block is 200 ft, that of an

---

[5] If there is really a bank at this corner, the similarity is purely coincidental.

east-and-west block 800 ft, and the average height of one floor of the Empire State building 12 ft, the three co-ordinate distances become 3200 ft in North-South direction, 400 ft in West-East direction, 936 ft in vertical direction. Using the Pythagorean

FIGURE 32

theorem we get now for the direct distance between two locations:

$$\sqrt{(3200)^2+(400)^2+(936)^2}=\sqrt{11.280.000}=3360 \text{ ft.}$$

If the concept of time as a fourth co-ordinate has any practical validity, we should now be able to combine the figure 3360 ft for space separation with the figure 15 min denoting the separation of the two events in time so as to obtain one single figure characterizing the *four-dimensional distance* between the two events.

According to the original idea of Einstein such *a four-dimensional distance can actually be determined by a simple generalization of the Pythagorean theorem and plays a more fundamental role in the physical relation between the events than do the individual space and time separations.*

FIGURE 33

Prof. Einstein was never able to do that. But he did something much better.

If we combine the space- and time-data, we must, of course, express them in comparable units just as it was necessary to designate in feet the lengths of blocks and the distance between floors. As we have seen above this can be done easily by using the velocity of light as the translation factor, so that the time interval of 15 min becomes 800,000,000,000 "light-feet." By a simple generalization of the Pythagorean theorem we should be inclined now to define the four-dimensional distance as the square root of the sum of the squares of all four co-ordinates: that is, three space and one time separation. In doing so we should, however, completely obliterate any difference between space and time, which would, in effect, be to admit the possibility of turning a space measurement into a time measurement and vice versa.

Yet nobody—not even the great Einstein—can, by covering a yardstick with a piece of cloth, waving a wand, and using some such magic phrase as: "pee-times-co-que-time-contra-variant-tensor," turn it into a brand new glittering alarm clock! (Figure 33.)

Thus, if we are going to identify time with space in the Pythagorean formula we must do it in some unconventional way that would preserve some of their natural differences.

According to Einstein, the physical difference between space distances and time durations can be emphasized in the mathematical formulation of a generalized Pythagorean theorem by using *the negative sign* in front of the square of the time co-ordinate. Thus we may designate the four-dimensional distance between two events as *the square root of the sum of the squares of the three space co-ordinates, minus the square of the time co-ordinate,* which has of course to be first expressed in space units.

The four-dimensional distance between the bank robbery and the plane crash is thus to be calculated as:

$$\sqrt{(3200)^2 + (400)^2 + (936)^2 - (800,000,000,000)^2}.$$

The exceedingly large numerical value of the fourth term as compared with the other three results from the fact that we took here an example from "ordinary life," and by ordinary life standards the rational unit of time is very small indeed. We should get more comparable figures if, instead of considering two events happening within New York City limits we were to take an example out of the cosmos. Thus taking as the first event the explosion of the atomic bomb at Bikini Atoll exactly at 9 A.M. on July 1, 1946, and as the second, say, the fall of a meteorite on the surface of Mars at 10 min after 9 A.M. the same day, we should have the time interval of 540,000,000,000 light-feet as compared with the space distance of about 650,000,000,000 ft.

In this case the four-dimensional distance between the two events would be: $\sqrt{(65\cdot10^{10})^2-(54\cdot10^{10})^2}$ ft$=36\cdot10^{10}$ ft, being numerically quite different in relation to both the pure-space and the pure-time intervals.

One may reasonably object, of course, to such a seemingly irrational geometry, in which one co-ordinate is treated differently from the other three, but one must not forget that any mathematical system devised to describe the physical world must be shaped so as to fit things, and if space and time do behave differently in their four-dimensional union, the laws of four-dimensional geometry must be shaped accordingly. Besides, there is a simple mathematical remedy that can make Einstein's geometry of space and time look exactly like the good old Euclidian geometry as we learned it in school. This remedy, proposed by the German mathematician Minkovskij, consists in considering the fourth co-ordinate as a purely imaginary quantity. You may remember from the second chapter of this book that one can turn an ordinary number into an imaginary one by multiplying it by $\sqrt{-1}$, and that such imaginary numbers can be used with great convenience in the solutions of various geometrical problems. Well, according to Minkovskij, in order to be considered as the fourth co-ordinate time must not only be expressed in space units but should also be multiplied by $\sqrt{-1}$. Thus the four co-ordinate distances pertaining to our example will be:

First co-ordinate:	3200 ft
Second co-ordinate:	400 ft
Third co-ordinate:	936 ft
Fourth co-ordinate:	$8\cdot10^{11}\times i$ light-feet.

We may now define the four-dimensional distance as the square root of the sum of the squares of *all four* co-ordinate distances. In fact, since the square of an imaginary number is always negative, the ordinary Pythagorean expression in Minkovskij's co-ordinates will be mathematically equivalent to the seemingly irrational Pythagorean expression in Einstein's co-ordinates.

There is a story about an old man with rheumatism who asked his healthy friend how he managed to avoid the malady.

"By taking a cold shower every morning all my life" was the answer.

"Oh!" exclaimed the first, "then you had cold showers *instead.*"

Well, if you do not like the seemingly rheumatic Pythagorean theorem, you can have the cold shower of the imaginary time co-ordinate instead.

The imaginary nature of the fourth co-ordinate in the space-time world, leads to the necessity of considering two physically different types of four-dimensional separations.

In fact, in such cases as the above discussed New York events, in which the three-dimensional distance between the events is numerically smaller than the time interval (in proper units), the expression under the radical in the Pythagorean theorem is negative so that we get an *imaginary number for the generalized four-dimensional separation.* In some other cases, however, the time duration is smaller than the space distance, so that we obtain a positive number under the radical. This means, of course, that in such cases *the four-dimensional separation between two events is real.*

Since, as discussed above, space-distances are to be considered as real whereas the time durations as purely imaginary, we may say that the real four-dimensional separations are related more closely to the ordinary space distances and the imaginary ones to the time intervals. According to *Minkovskij's* terminology, the four-dimensional separations of the first kind are called *spatial* (raumartig) and those of the second *temporal* (zeitartig).

We shall see in the next section that the spatial separation can be turned into a regular space distance, and the temporal separation into a regular time interval. However, the fact that one of them is represented by a real number whereas the other is represented by an imaginary number forms an insurmountable barrier in any attempt to turn one into another, making it impossible for us to turn a yardstick into a clock or a clock into a yardstick.

CHAPTER V

Relativity of Space and Time

1. *TURNING SPACE INTO TIME AND VICE VERSA*

ALTHOUGH mathematical attempts to demonstrate the unity of space and time in a single four-dimensional world do not completely obliterate the differences between distances and durations, they certainly reveal a much greater similarity between the two notions than was ever evident in pre-Einsteinian physics. In

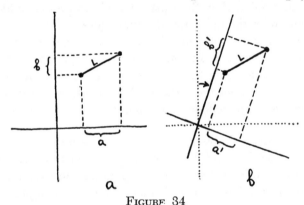

FIGURE 34

fact, *space distances and time intervals between various events must now be considered only as the projections of the basic four-dimensional separation between these events on the space and on the time axis, so that the rotation of the four-dimensional axis-cross may result in partial transformation of distances into durations and vice versa.* But what do we mean by the rotation of the four-dimensional space-time axis-cross?

Let us first consider an axis-cross made by the two space-co-ordinates as shown in Figure 34a, and suppose we have two fixed points separated by a certain distance *L*. Projecting this distance on the co-ordinate axis, we find that our two points are separated by *a* ft in the direction of the first axis, and by *b* ft in the direction of the second. If we turn the axis-cross by a certain angle (Figure 34*b*) the projections of the same distances on the two new axes will be different from the previous projections, possessing the new values *a′* and *b′*. However, according to the Pythagorean theorem, the square root of the sum of the squares of the two projections will be the same in both cases since it corresponds to the *actual* distance between the points, which does not change because of axis rotation. Thus

$$\sqrt{a^2+b^2}=\sqrt{a'^2+b'^2}=L.$$

We say that the square root of the sum of the squares is invariant in respect to the rotation of co-ordinates, whereas the

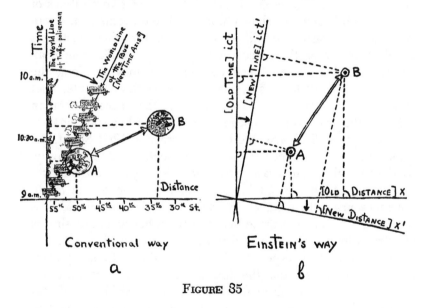

Conventional way   Einstein's way

a   b

FIGURE 35

particular values of the projections are incidental and depend on the choice of co-ordinate systems.

Let us consider now the axis-cross in which one axis corresponds to a distance and another to a duration. In this case the two fixed points of the previous example become the two fixed events, and the projections on the two axes represent respectively their separations in space and in time. Taking for the two events the bank robbery and the plane crash discussed in the previous section, we can draw a picture (Figure 35a) that is very similar to that representing the two space co-ordinates (Figure 34a). What must we do now in order to turn the axis-cross? The answer is rather unexpected and even perplexing: If you want to turn the space-time axis-cross, get on a bus.

Well, suppose we really sit on the upper deck of a bus going down Fifth Avenue on the fatal morning of July 28. From our own egoistic point of view, we shall in this case be mostly interested in the question of *how far away from our bus* the bank robbery and the plane crash take place, if only because the distances determine whether or not we could see what was happening.

If you look at Figure 35a, in which the successive positions of the bus's world-line are shown along with the events of the robbery and the crash, you will notice at once that these distances are different from those recorded by, say, a traffic policeman standing on his corner. Since the bus was moving along the avenue, advancing, let us say, one block every three minutes (not so unusual in heavy New York traffic!), the space separation between the two events as seen from the bus becomes smaller. In fact, since at 9:21 A.M. the bus was crossing 52nd Street, the bank robbery, which occurred at this moment, was 2 blocks away. By the time the plane crash took place (9:36 A.M.) the bus was at 47th Street, that is, 14 blocks from the scene of the crash. Thus measuring the distances *relative to the bus* we should conclude that the space distance between the robbery and crash was $14 - 2 = 12$ blocks, as compared with the distance of $50 - 34 = 16$ blocks measured in respect to the city buildings. Looking again at Figure 35a, we see that the distances as recorded from the bus must be counted not from the vertical axis (world line of stationary policeman) as before, but rather from the inclined line representing the world-line of the bus, so that it is this latter line that is now playing the role of the new time axis.

The "pack of trivialities" just discussed may be summarized in this statement: to plot the space-time diagram of events as they were observed from a moving vehicle, we must turn the time axis by a certain angle (depending on the velocity of that vehicle), leaving however the space axis intact.

This statement, although a gospel truth from the point of view of classical physics and so-called "common sense," stands, however, in direct contradiction to our new ideas concerning the four-dimensional space-time world. If, in fact, time is to be considered as the independent fourth co-ordinate, *the time axis must always remain perpendicular to the three space axes,* regardless of whether we sit on a bus, a trolley, or on the pavement!

At this point we may follow either of two paths of thought. Either we have to retain our conventional ideas of space and time, abandoning any further consideration of the unified space-time geometry, or we must break with the old ideas dictated by "common sense," and assume that in our space-time diagram *the space axis must be turned along with the time axis,* so that the two always remain mutually perpendicular (Figure 35*b*).

But, in the same way that turning the time axis means physically that *the space separation of two events has different values* (12 and 16 blocks in the previous example) *when viewed from a moving vehicle,* turning the space axis would mean that *the time separation of two events observed from a moving vehicle differs from the time separation of the two events when observed from a stationary point on the ground.* Thus, if the bank robbery and plane crash were 15 minutes apart by the City Hall clock, the time interval registered by the wrist watch of a bus passenger would be different—not because the two timepieces move at different rates as the result of mechanical imperfections, but because *time itself flows at different rates in vehicles moving at different speeds, and the actual mechanism that records it is correspondingly slowed,* though, at the low speeds of bus travel, this retardation is so negligible as to be imperceptible. (This phenomenon will be discussed at greater length in this chapter.)

To give one more example, let us consider a man eating his dinner in the dining car of a moving train. From the point of view of the dining-car waiter he eats his appetizer and his dessert at the very same place (third table near the window). But from the point of view of two switchmen at stationary points on the railroad track looking through the window of the car—the one just in time to see him eat his appetizer, the other just in time to see him eat his dessert—the two events take place many miles apart. Thus we may say that: *Two events occurring at the same place, but at two different moments from the point of view of one observer, will be considered as occurring at different places if viewed by other observers in a different state, or in different states, of motion.*

In view of the desired space-time equivalence, replace in the above sentence the word "place" by the word "moment" and vice versa. The sentence will now read: *The two events occurring at the same moment, but at different places, from the point of view of one observer, will be considered as occurring at different moments if viewed by another observer in a different state of motion.*

In application to our dining-car example, we would expect that whereas the waiter would swear that two passengers sitting at opposite ends of the car lighted their after-dinner cigarettes at exactly the same moment, a switchman standing still on the track and looking through the windows as the train moved past him would insist that one of these gentlemen had done it before the other.

Thus: *two events considered to be simultaneous from the point of view of one observer will from the point of view of another be separated by a certain time interval.*

These are the inevitable consequences of the four-dimensional geometry in which *space and time are only the projections of an invariant four-dimensional separation on corresponding axes.*

2. *ETHER WIND, AND SIRIUS TRIP*

Let us now ask ourselves whether the mere desire to use the language of four-dimensional geometry justifies the introduction of such revolutionary changes into our old and comfortable ideas about space and time?

If our answer is yes, we challenge the entire system of classical physics, which is based on the definitions of space and time formulated by the great Isaac Newton two and a half centuries ago: "Absolute space, in its own nature, without relation to anything external, remains always similar and immovable," and "Absolute, true, and mathematical time, of itself, and from its

own nature, flows equally without relation to anything external."
In writing these lines Newton certainly did not think that he was
stating anything new, or anything open to argument; he was
simply formulating in an exact language the notions of space and
time as they were apparent to anybody with common sense. In
fact the belief in the correctness of these classical ideas about
space and time was so absolute that they have been often held
by philosophers as *a priori,* and no scientist (not to mention
laymen) ever considered the possibility that they might be false,
and thus in need of re-examination and restatement. Why, then,
should we reconsider the question now?

The answer is that the abandonment of classical ideas of space
and time and their unification in a single four-dimensional picture were dictated not by any purely esthetic desire on the part
of Einstein, nor by any mere restless urge of his mathematical
genius, but by stubborn facts that emerged constantly from
experimental research, and that just wouldn't fit into the classical
picture of independent space and time.

The first impact against the very foundations of the beautiful
and, apparently eternal, castle of classical physics, an impact that
shook practically every single stone of this elaborate building
and sent its walls tumbling down, like the walls of Jericho before
the blast of Joshua's trumpet, was delivered by what would seem
to be an unpretentious experiment carried out in 1887 by an
American physicist, A. A. Michelson. The idea of Michelson's experiment is very simple and is based on a physical picture according to which light represents some kind of wave motion traveling
through the so called "light-carrying ether," a hypothetical substance uniformly filling up interstellar space as well as the intervals between the atoms in all material bodies.

Drop a stone into a pond, and waves will ripple out in all directions. The light that comes from any bright body similarly ripples
out in waves, and so does the sound of a vibrating tuning fork.
But, whereas the surface waves clearly represent the motion of
the particles of water, and the sound waves are known to be the
vibration of the air or other materials through which sound is
traveling, we are unable to find any material medium that is
responsible for carrying light waves. In fact, the space through

which light travels with such ease (in contrast to sound) seems to be completely empty!

Since, however, it seems rather illogical to speak about something vibrating when there is nothing to vibrate, physicists had to introduce a new notion, "light-carrying ether," in order to furnish a substantive subject for the verb "to vibrate" when attempting to explain the propagation of light. From the purely grammatical point of view, which requires that any verb must necessarily have a subject, the existence of the "light-carrying ether" cannot possibly be denied. But—and it is a very large "but"—the rules of grammar do not, and cannot, prescribe to us the physical properties of the substantives that must be introduced in a correctly constructed sentence!

If we say that light consists of waves traveling through the light ether, defining "light ether" as *that* through which light waves are traveling, we are telling a gospel truth, but also recording a most trivial tautology. It is an entirely different problem to find out *what* this light ether is and what its physical properties are. Here no grammar (not even Greek!) can help us, and the answer must come from the science of physics.

As we shall see in the course of the following discussion, the greatest mistake of the physics of the nineteenth century consisted in the assumption that this light ether has properties very similar to those of ordinary physical substances familiar to us. One used to speak about the fluidity, rigidity, various elastic properties, and even the internal friction of light ether. Thus, for example, the fact that light ether behaves on the one hand as a vibrating solid when carrying light waves,[6] but on the other hand shows a perfect fluidity and a complete absence of any resistance to the motion of celestial bodies, was interpreted by comparing it with such materials as sealing wax. Sealing wax, and other similar substances, are, in fact, known to be quite hard and brittle in respect to forces acting rapidly in a mechanical impact, but will flow like honey under the force of their own

[6] With respect to light waves the vibrations were shown to be transverse to the direction in which light was traveling. In ordinary materials such transverse vibrations occur only in solids, whereas in liquid and gaseous substances vibrating particles can move only in the direction in which the wave is proceeding.

weight if left alone for a sufficiently long time. Following this analogy, the old physics assumed that light ether, filling all interstellar space, acted as a hard solid in respect to very rapid distortions connected with the propagation of light, but behaved as a good liquid when the planets and stars, moving many thousand times slower than light, were pushing their way through it.

Such an anthropomorphical point of view, so to speak, which tried to ascribe to a completely unknown thing, which so far had nothing but the name, the properties of ordinary material known to us, failed very badly from the very beginning. And, in spite of many attempts, no reasonable mechanical interpretation of the properties of the mysterious carrier of light waves was found possible.

In the light of our present knowledge we can easily see wherein all attempts of that kind erred. In fact we know that all mechanical properties of ordinary substances can be traced back to the interaction between the atoms from which they are built. Thus, for example, the high fluidity of water, the elasticity of rubber, and the hardness of a diamond depend on the fact that water molecules can slide by each other without much friction, that rubber molecules can be easily deformed, and that the atoms of carbon forming a diamond crystal are tightly bound together into a rigid lattice. Thus all common mechanical properties of various substances result from their atomic structure, but this rule makes no sense whatsoever when applied to an absolutely continuous substance such as that which light ether is considered to be.

Light ether is a substance of a peculiar type, which has no similarity to the familiar atomic-mosaic that we usually call matter. We can call light ether a "substance" (if only because it serves as a grammatical subject for the verb "to vibrate"), but we can also call it "space," keeping in mind that, as we have seen before and will see again, space may possess certain morphological or structural features that make it a much more complicated thing than it is in the conceptions of Euclidian geometry. In fact, in modern physics the expressions "light ether" (divested of its alleged mechanical properties) and "physical space" are considered synonymous.

But we have deviated too far into the gnosiological or philosophical analysis of "light ether," and must return now to the subject of Michelson's experiment. As we have said before, the idea of this experiment is quite simple. If light represents the waves traveling through ether, the velocity of light as recorded by instruments located on the surface of the earth must be distorted by the motion of the earth through space. Standing on the earth which rushes along its orbit around the sun, we should experience an "ether wind," in the same way that a man on the deck of a fast moving ship feels the wind blowing into his face though the weather may be perfectly calm. Of course we do not feel the "ether wind," since it is supposed to penetrate without any difficulty between the atoms forming our body, but we should be able to detect its presence by measuring the velocity of light in different directions in relation to our motion. Everybody understands that the velocity of a sound traveling in the same direction as the wind is greater than that of the same sound traveling against the wind, and it seems natural that the same thing should be true of light propagating with and against the ether wind.

Reasoning thus, Professor Michelson set out to construct an apparatus that could register the differences in the speed of light propagating in different directions. The simplest way to accomplish it would be, of course, to take the apparatus of Fizeau described above (Figure 31C), and to perform a series of measurements, turning it in different directions. This would not, however, be a very rational way of doing it, because it would require a high degree of precision in each case. Indeed since the expected difference (equal to the velocity of the earth) is only about one hundredth of one per cent of the speed of light, we should have to perform each individual measurement with extremely great accuracy.

If you have two long sticks of about the same length, and want to know exactly the difference between them, you will find the difference most easily by putting them together at one end and measuring the difference at the other end. This is known as the "zero-point" method.

Michelson's apparatus, shown schematically in Figure 36, utilizes this zero-point method for comparing the velocities of light in two perpendicular directions to each other.

The centerpiece of this apparatus is formed by a glass plate *B* covered with a thin semitransparent layer of silver, which reflects about 50 per cent of the incident light and lets through the other

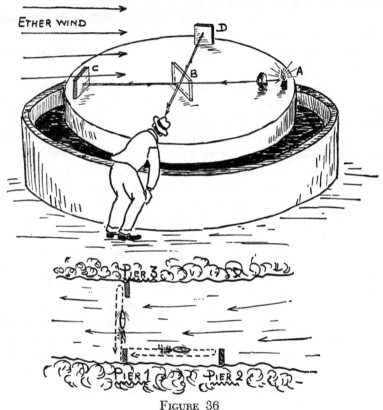

FIGURE 36

50 per cent. Thus the light beam coming from the source **A** is split into two equal parts traveling parallel to each other. These two beams are reflected from the two mirrors *C* and *D* placed at equal distances from the central plate, and are sent back to it.

The beam coming back from D will be partially transmitted by the thin silver layer, and will unite with the part of the beam from C that is partially reflected by the same layer. Thus the two beams separated at the entrance to the apparatus will reunite as they enter the observer's eye. According to a well-known law of optics, the two beams will interfere with each other, forming a system of dark and light fringes visible to the eye.[7] If the distances BD and BC are equal, so that the two beams return to the centerpiece simultaneously, the bright fringe will be in the center of the picture. If the distances are slightly changed so that one beam is delayed in respect to the other the fringes will be shifted to the right or to the left.

Since the apparatus is placed on the surface of the earth and since the earth moves rapidly through space, we must expect that the ether wind is blowing through it with a speed equal to the speed of the earth motion. Assume, for example, that this wind is blowing in the direction from C to B (as shown in Figure 36), and let us ask ourselves what difference it makes in the speed of the two beams hurrying to their meeting point.

Remember that one of these beams goes first against the wind and returns with it, whereas another beam goes across the wind both ways. Which will return first?

Think of a river, and a motorboat proceeding upstream from Pier 1 to Pier 2 and then returning downstream to Pier 1. The stream hampers it on the first part of the journey, but helps its motion on the way back. You may be inclined to believe that the two effects compensate each other, but this is not so. In order to understand this, imagine that the boat goes with a speed equal to the speed of the stream. In this case the boat from 1 will never be able to reach Pier 2! It is not difficult to see that the presence of the stream will in all cases lengthen the time of the round trip by a factor of

$$\frac{1}{1-\left(\dfrac{V}{v}\right)^2}$$

where v is the velocity of the boat and V the velocity of the

[7] See also pp. 122-23.

stream.[8] Thus, for example, if the boat travels ten times faster than the stream, the return trip will last:

$$\frac{1}{1-\left(\frac{1}{10}\right)^2} = \frac{1}{1-0.01} = \frac{1}{0.99} = 1.01 \text{ times,}$$

that is, 1 per cent longer than it would in quiet water.

In a similar way we can also calculate the expected delay of the round trip across the river. Here the delay arises from the fact that, in order to reach Pier 3 from Pier 1, the boat must travel slightly sidewise to compensate for the drift in the moving water. In this case the delay is somewhat less. being indicated by the factor:

$$\sqrt{\frac{1}{1-\left(\frac{V}{v}\right)^2}}$$

that is, by only 1/2 per cent for the example above. The proof of this formula is very simple and we leave it to the inquisitive reader. Now, for the river substitute streaming ether, for the boat substitute the light wave propagating through it, and for the piers substitute the two end mirrors, and you will have the scheme of Michelson's experiment. The beam of light going from B to C and returning to B will now be delayed by a factor of

$$\frac{1}{1-\left(\frac{V}{c}\right)^2}$$

c being the speed of light through the ether, whereas the light traveling from B to D and back must be delayed by the factor

$$\sqrt{\frac{1}{1-\left(\frac{v}{c}\right)^2}}$$

[8] In fact writing l for the distance between the two piers, and remembering that the combined speed downstream is $v+V$ and upstream $v-V$ we obtain for the total time of the round trip:

$$t = \frac{l}{v+V} + \frac{l}{v-V} = \frac{2vl}{(v+V)(v-V)} = \frac{2vl}{v^2-V^2} = \frac{2l}{v} \cdot \frac{1}{1-\dfrac{V^2}{v^2}}$$

Since the velocity of ether wind, which is equal to the velocity of the Earth, is 30 km per second, and the velocity of light is 3×10^5 km/sec the two beams must be delayed respectively 0.01 and 0.005 per cent. Thus it should be a simple matter to observe, with the aid of Michelson's apparatus, the difference in the speed of a beam of light traveling with the ether wind, and that of one traveling against it.

You may imagine Michelson's surprise, then, when, in performing the experiment, he was unable to notice even the slightest shift of the interference fringes.

Apparently the ether wind had no effect on the velocity of light whether it was traveling along or across it.

The fact was so astonishing that Michelson himself did not believe it at first, but careful repetitions of the experiment left no doubt that, astonishing as it was, the result he had obtained at first was correct.

The only possible explanation of this unexpected result seemed to lie in the bold assumption that the massive stone table on which Michelson's mirrors were mounted contracted slightly (the so called Fitz-Gerald[9] contraction) in the direction of the earth's motion through space. In fact, if the distance BC shrinks by a factor of

$$\sqrt{1-\frac{v^2}{c^2}}$$

whereas the distance BD remains unaltered, the delay of both light beams becomes equal and no shift of interference fringes is to be expected.

But it was easier to suggest the possibility that Michelson's table had shrunk than to understand it. True, we do expect some contraction of material bodies moving through a resisting medium. A motorboat racing across a lake, for example, is slightly squeezed between the driving force of the propeller at its stern and the water resistance at the bow. But the extent of such mechanical contraction depends on the strength of the material from which the boat is made. A steel boat will be squeezed by a smaller degree than a wooden one. But variations in the con-

[9] Named for the physicist who first introduced that notion considering it as a purely mechanical effect of motion.

traction that was responsible for the negative results in Michelson's experiment depends only on the speed of motion and not at all on the strength of the materials involved. Had the table mounting the mirrors been made not from stone, but from cast iron, wood, or any other material, the amount of contraction would have been exactly the same. It is thus clear that we deal here with a *universal effect*, which causes all moving bodies to contract in exactly the same degree. Or, to describe the phenomenon as Professor Einstein did in 1904, *we deal here with the contraction of space itself, and all material bodies moving with*

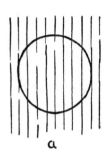 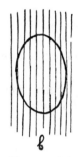 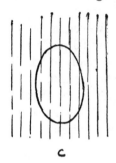

a   b   c

FIGURE 37

the same speed contract in the same way simply because they are imbedded in the same contracted space.

In the course of the last two chapters we have said enough about the properties of space to make the above statement sound reasonable. In order to make the situation clearer we can imagine that space has some properties of an elastic jelly in which the boundaries of different bodies are traced. When space is distorted by being squeezed, stretched, or twisted, the shapes of all bodies imbedded in it change automatically in the same way. These distortions of material bodies caused by the distortion of space must be distinguished from individual distortions caused by various external forces that produce internal stresses and strains in the bodies so distorted. Examination of Figure 37 representing a two-dimensional case will probably help to explain this important difference.

However, the effect of space shrinking, though it is of fundamental importance in understanding the basic principles of physics, passes quite unnoticed in ordinary life, since the highest velocities that affect us in our everyday experience are still negligibly small as compared with the velocity of light. Thus, for example, a car speeding at 50 miles an hour is reduced in length by a factor of $\sqrt{1-(10^{-7})^2} = 0.99999999999999$, which corresponds to a reduction in bumper-to-bumper length by only *the diameter of one atomic nucleus!* A jet-propelled plane flying at a speed of over 600 miles an hour is reduced in length by only one atomic diameter, and an interstellar rocket 100 m long rushing at a speed of over 25,000 miles an hour, by one hundredth of a millimeter.

However, if we can imagine objects moving with speeds 50, 90, and 99 per cent of light speed, their lengths will be reduced respectively to 86, 45, and 14 per cent of their sizes when standing on the ground.

This effect of relativistic contraction of all fast-moving objects is commemorated in the following limerick written by an unknown author:

> "There was a young fellow named Fisk
> Whose fencing was exceedingly brisk.
> So fast was his action,
> The Fitz-Gerald contraction
> Reduced his rapier to a disk."

This Mr. Fisk must have been fencing with lightning speed indeed!

From the point of view of four-dimensional geometry the observed universal shortening of all moving objects can be simply interpreted as the change of the space projection of their invariant four-dimensional length caused by the rotation of the space-time axis-cross. In fact, you must remember from the discussion in the previous section that the observations carried from a moving system must be described by means of co-ordinates in which the space and time axis are both turned by a certain angle depending on the velocity. Thus if in the resting system we had a certain four-dimensional separation projecting a hundred per cent on the space axis (Figure 38*a*), its space projection on the new time axis (Figure 38*b*) would always be shorter.

The important point to remember is that the expected shortening of length is entirely relative with respect to the two systems moving with respect to each other. If we consider an object that is at rest with respect to the second system, thus being represented by an invariant line parallel to the new space axis, its projection on the old axis will be shortened by the same factor.

Thus there is no necessity, and in fact no physical sense, in specifying *which* of the two systems is "actually" in motion. All that matters is only that they are in motion in relation to each other. Thus if two passenger rocket ships of some future "Interplanetary Communication Co. Ltd." were to meet somewhere in space between Earth and Saturn, traveling at very high speed,

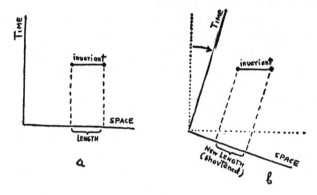

Figure 38

the passengers of each ship would be able to see through the side windows that the other ship is considerably shrunk, whereas they would notice no shrinking of their own ship. And it would be quite useless to argue which ship is "actually" shrunk, since each is from the point of view of the passengers on the other ship, and neither is from the point of view of its own passengers.[10]

The four-dimensional reasoning also permits us to understand why the relativistic shrinkage of moving objects becomes appreciable only when their velocity approaches that of light. In fact,

[10] Of course this is all a theoretical picture. Actually if two rocket ships passed each other traveling at such speeds as we are here considering, the passengers on either ship would not be able to see the other at all—any more than you can see a bullet fired from a rifle at a fraction of this speed.

the angle by which the space-time axis-cross is turned is determined by the ratio of the distance covered by the moving system to the time necessary to cover this distance. If we measure the distances in feet and time in seconds this ratio would be nothing else but the ordinary velocity expressed in feet per second. Since, however, the time-intervals in the four-dimensional world are represented by the ordinary time-interval multiplied by the velocity of light, the ratio determining the rotation angle is actually the velocity of motion in feet per seconds divided by the velocity of light in the same units. Thus the angle of rotation, and its influence on distance measurements becomes appreciable only when the relative velocity of the two moving systems approaches the velocity of light.

In the same way as it influences the length measurements, the turning of the space-time axis-cross affects the measurements of time intervals. One can show, however, that because of the peculiar imaginary nature of the fourth co-ordinate,[11] the time intervals will expand when the space distances shrink. If you have a clock mounted on a fast moving car it will go somewhat slower than a similar clock on the ground, so that the time interval between two successive ticks will be lengthened. Just as in the case of shortening of lengths, the slowing down of a moving clock is a universal effect depending only on the velocity of motion. The modern wrist watch, the old-fashioned grandfather's clock with a pendulum, or the hour glass with running sand will be slowed down in exactly the same way provided they move with the same velocity. The effect is, of course, not limited to special mechanical gadgets that we call "clocks" and "watches"; in fact, all physical, chemical, or biological processes will be slowed down in the same degree. Thus there is no danger that when cooking the eggs for breakfast in a fast moving rocket ship you will overcook them because your watch runs too slowly; the processes inside of the egg will be slowed correspondingly so that keeping them in boiling water for five minutes according to your watch, you will get what you have always known as "five-

[11] Or, if you wish, because of the fact that the Pythagorean formula in the four-dimensional space is distorted in respect to time.

minute eggs." We use a rocket ship here as an example rather than a train's dining car, because, as in the case of length contraction, the expansion of time becomes noticeable only at velocities approaching the speed of light. This time expansion is given by the same factor

$$\sqrt{1-\frac{v^2}{c^2}}$$

as the space-contraction, with the difference that here you use it not as a multiplier but as a divisor; if one moves so fast that the lengths are reduced to one half, the time intervals become twice as long.

The slowing down of the speed of time in moving systems has an interesting implication in respect to interstellar travel. Suppose you decided to visit one of the satellites of Sirius, which is at a distance of nine light-years from the solar system, and use for your trip a rocket ship that can move practically with the speed of light. It would be natural for you to think that the round trip to Sirius and back would take you at least eighteen years, and you would be inclined to take with you a very large food supply. That precaution, however, would be absolutely unnecessary if the mechanism of your rocket ship made it possible for you to travel at nearly the velocity of light. In fact if you move, for example, at 99.99999999 per cent of the speed of light, your wrist watch, your heart, your lungs, your digestion, and your mental processes will be slowed down by a factor of 70,000, and the 18 yrs (from the point of view of people left on the Earth) necessary to cover the distance from Earth to Sirius and back to Earth again, would seem to you as only a few hours. In fact, starting from Earth right after breakfast, you will just feel ready for lunch when your ship lands on one of the Sirius planets. If you are in a hurry, and start home right after lunch, you will, in all probability, be back on Earth in time for dinner. But, and here you will get a big surprise if you have forgotten the laws of relativity, you will find on arriving home that your friends and relatives have given you up as lost in the interstellar spaces and have eaten 6570 dinners without you! Because you were traveling at

a speed close to that of light, 18 terrestrial years have appeared
to you as just 1 day.

But what about trying to move faster than light? The answer
to this question can be partially found in another relativistic
limerick:

> "There was a young girl named Miss Bright,
> Who could travel much faster than light.
> She departed one day,
> In an Einsteinian way,
> And came back on the previous night."

To be sure, if speeds that approach the velocity of light make
time in a moving system run slower, a superlight velocity should
turn the time backward! Besides, owing to the change of the
algebraic sign under the Pythagorean radical, the time co-ordinate would become real and thus indicate a distance in space in
the same way all lengths in the superlight-speed system go
through zero and become imaginary, thus turning into time-
intervals.

If all this were possible, the picture in Figure 33 showing Einstein turning a yardstick into an alarm clock would correspond
to reality provided that during this performance he could assume
a superlight speed!

But the physical world, crazy as it is, is not *that* crazy, and the
obvious impossibility of such a black-magic performance can be
simply summarized by the statement that *no material object can
move with a speed that equals or exceeds the speed of light.*

The physical foundation for this basic law of nature lies in the
fact, proved by numerous direct experiments, that *the so-called
inertial mass of moving objects, which measures their mechanical
resistance to further acceleration, increases beyond any limit
when the velocity of motion approaches that of light.* Thus if a
revolver bullet moves with a speed 99.99999999 per cent of light
speed its resistance to further acceleration is equivalent to that
of a 12-in gun shell. And at the speed of 99.99999999999999 per
cent of light speed, our little bullet will have the same inertial
resistance as a heavily loaded freight car. Regardless of how
great an effort we applied to our bullet, we should never be able

to conquer the last decimal and to make its speed exactly equal the upper speed limit for all motion in the universe!

3. CURVED SPACE, AND THE RIDDLE OF GRAVITY

With due repentance and apologies to the poor reader, who must feel as though he had been stumbling over all four co-ordinate axes in the course of the last twenty pages, we now invite him

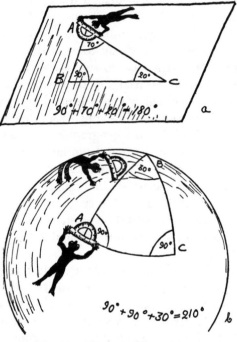

FIGURE 39

Two-dimensional scientists of the flat and curved "surface worlds" check the Euclidian theorem about the sum of the angles in a triangle.

to come for a walk in a *curved space*. Everybody knows what a curved line and a curved surface are, but what could one possibly mean by the expression "a curved space"? The difficulty encountered in trying to imagine such a phenomenon lies, not so much in the unusualness of the conception, as in the fact that, whereas we can look at curved lines and curved surfaces from the outside, the curvature of three-dimensional space must be observed *from inside* since we ourselves are within it. In an attempt to understand how a three-dimensional human being can conceive the curvature of the space in which he lives, let us first consider the hypothetical situation of two-dimensional shadow beings living on a surface. In Figures 39a and 39b we see the shadow scientists of flat and curved (spherical) "surface-worlds" studying the geometry of their two-dimensional spaces. The simplest geometrical figure to study is, of course, a triangle, that is, the figure formed by three straight lines connecting three geometrical points. As everybody will remember from high school geometry, the sum of the three angles of any triangle drawn on a plane is always equal to 180°. It is easy to see, however, that the above theorem does not apply to triangles drawn on the surface of a sphere. Indeed, a spherical triangle formed by the sections of two geographical meridians diverging from the pole, and the section of the parallel (also in a geographical sense) cut by them, has two right angles at the base and can have any angle between zero and 360° at the top. In the particular example that is being studied by two shadow scientists in Figure 39b, the sum of the three angles equals 210°. Thus we see that, by measuring geometrical figures in their two-dimensional world, the shadow scientists can discover its curvature without actually looking at it from the outside.

Applying the above observations to a world that has one more dimension we quite naturally come to the conclusion that *human scientists living in three-dimensional space can ascertain the curvature of that space without jumping out into the fourth dimension simply by measuring the angles between the straight lines connecting three points in their space.* If the sum of the three angles equals 180° the space is flat; otherwise the space must be curved.

But before we carry the argument further, we must discuss in some detail exactly what is meant by the expression *straight line.* Looking at the two triangles shown in Figures 39a and 39b the reader would probably say that, whereas the sides of the triangle on the plane (Figure 39a) are truly straight lines, the sides of that on the sphere (Figure 39*b*) are actually curved, being the arcs of great circles[12] that conform to the spherical surface.

Such a statement, based on our common-sense geometrical ideas, would deny to the shadow scientists any possibility of developing the geometry of their two-dimensional space. The concept of the straight line needs a general mathematical definition that not only will hold in place Euclidian geometry, but also could be extended to include lines on the surfaces and in spaces of a more complicated nature. Such a generalization can be obtained by *defining a "straight line" as the line representing the shortest distance between two points, conforming to the surface or the space within which it is drawn.* In plane geometry the above definition coincides, of course, with the common concept of a straight line, while in more complicated cases of curved surfaces it leads to a well-defined family of lines, which play here the same role as that of the ordinary "straight lines" in the geometry of Euclid. To avoid misunderstanding, one often calls the lines representing the shortest distances on curved surfaces *geodesical lines* or *geodesics*, because this notion was first introduced in *geodesy*, that is, the science of measurements on the surface of the Earth. In fact, when we speak of the straight-line distance between New York and San Francisco, we mean "straight as the crow flies" following the curve of the earth's surface and not as a hypothetical gigantic miner's drill would advance, boring its way straight through the body of the earth.

The above definition of the "generalized straight line" or "geodesic" as the shortest distance between two points suggests the simple physical method of constructing such lines by *stretching a piece of string between the points in question.* If you do it on a plane surface, you will describe an ordinary straight line; doing it on a sphere you will find that the string stretches along the arc of a great circle, which corresponds to the geodesic of the spherical surface.

In a similar way it should be possible to find out whether the three-dimensional space in which we live is flat or curved. All

[12] Great circles are the circles cut on the surface by a plane passing through the center of the sphere. Equator and meridians are such great circles.

that is necessary is to stretch strings between three points in
space, and to see whether the sum of the angles thus formed is
or is not equal to 180°. In planning such an experiment we must
remember, however, two important points. It is essential that the
experiment be done on a rather large scale since a very small
part of the curved surface or space may appear to us quite flat;

FIGURE 40A

obviously we cannot ascertain the curvature of the Earth's surface by measurements made in our backyards! Further, the
surface or the space may be flat in some regions, and curved in
others, so that a complete survey may be necessary.

The great idea, which was included by Einstein in the foundation of his general theory of curved space, consists of the assumption that *the physical space becomes curved in the neighborhood
of large masses*; the bigger the mass the larger the curvature.
In an attempt to verify such a hypothesis experimentally we
might stretch a string between three spikes driven into the ground
around some nice big hill (Figure 40A), and measure the angles
formed by the strings at their three points of meeting. Choose
the biggest hill you can find—even one of the Himalaya Mountains—and you will find that, making allowances for possible
errors in your measurements, the sum of the three angles where

the strings meet will be exactly 180°. However, this result would not necessarily mean that Einstein is wrong, and that the presence of big masses does not curve the space around them. Perhaps even the Himalaya Mountains do not make the surrounding space curve enough so that the deviation can be recorded by even our most precise measuring instruments; remember the fiasco encoun-

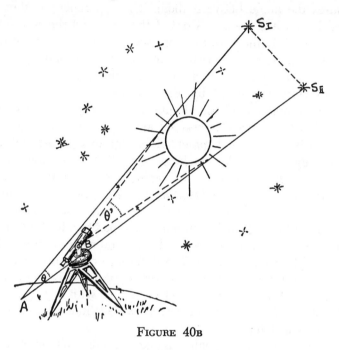

FIGURE 40B

tered by Galileo in his attempt to measure the speed of light by means of his shutter-lanterns! (Figure 31).

So you must not be discouraged, but must try again with a still bigger mass, the sun for example.

And lo, here is success! You will find, if you stretch a string from some point on the earth, to some star, then to another star, then back to the original point on earth, choosing the stars so that the sun is corralled in the triangular enclosure formed by the strings, the sum of the three angles will be *noticeably* different from 180°. If you have not a string long enough for such an experiment, substitute for the string a ray of light, which is every bit as good, since optics teach us that light always takes *the shortest possible route*.

Such an experiment in measuring angles formed by the beams of light is represented schematically in Figure 40B. The light rays from two stars S_I and S_{II} located (at the moment of observation) at opposite sides of the sun disc converge into a theodolite, which measures the angle between them. The experiment is then repeated later when the sun is out of the way, and the two angles are compared. If they are different we have proof that the mass of the sun changes the curvature of the space around it, deflecting the rays of light from their original paths. Such an experiment was originally suggested by Einstein to test his theory. The reader may understand the situation somewhat better by comparing it with its two-dimensional analogy shown in Figure 41.

Obviously there was a practical barrier to carrying out Einstein's suggestion under ordinary conditions: because of the brilliance of the solar disc, you cannot see the stars around it; but during a total solar eclipse the stars are clearly visible in the daytime. Taking advantage of this fact, the test was actually made in 1919 by a British astronomical expedition to the Principe Islands (West Africa), from which the total solar eclipse of that year could best be observed. The difference of angular distances between the two stars with and without the sun between them was found to be $1.61'' \pm 0.30''$ as compared with 1.75 predicted by Einstein's theory. Similar results were obtained by various expeditions at later dates.

Of course, one and a half angular seconds isn't much of an angle, but it is enough to prove that the mass of the sun *does* force the space around it into a curve.

If, instead of the sun, we could use some other much bigger star, the Euclidian theorem about the sum of the three angles in a triangle would be found wrong by angular minutes or even by degrees.

It takes some time and a great deal of imagination to become accustomed to the notion of the curved three-dimensional space as viewed by an inside observer, but once you get it right, it will stand out as clearly and as definitely as any other familiar concept of classical geometry.

We need now to make only one more important step in order to understand completely Einstein's theory of curved space and its relation to the fundamental problem of universal gravitation. To do this we must remember that the three-dimensional space that we have been discussing represents only a part of the four-dimensional space-time world that serves as the background for all physical phenomena. Thus the curvature of space proper must be only the reflection of the more general four-dimensional curvature of the space-time world, and *the four-dimensional world-lines representing the motion of light rays and material objects*

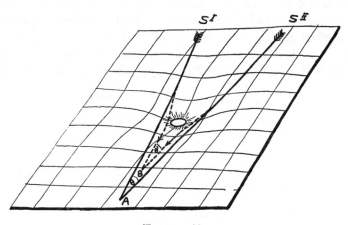

FIGURE 41

in this world must be looked upon as the curved lines in the superspace.

Examining the question from this point of view, Einstein came to the remarkable conclusion that *the phenomenon of gravity is merely the effect of the curvature of the four-dimensional space-time world.* In fact, we may now discard as inadequate the old statement that the sun exercises a certain force that acts directly on the planets making them describe circular orbits around it. It would be more accurate to say that *the mass of the sun curves the space-time world around it, and that the world-line of planets look the way they do in Figure 30 only because they are geodesical lines running through the curved space.*

Thus the concept of gravity as an independent force completely disappears from our reasoning, and is replaced by concepts of the pure geometry of space in which all material objects move along the "straightest lines," or geodesics, following the curvature produced by the presence of other big masses.

4. *CLOSED AND OPENED SPACES*

We cannot conclude this chapter without a brief discussion of another important problem of Einstein's space-time geometry: the dilemma of the finite and the infinite universe.

So far we have been discussing the local curvature of space in the neighborhood of large masses, a variety of "space pimples" scattered over the giant face of the universe. But, apart from these local deviations, is the face of the universe flat or is it curved, and if so which way? In Figure 42 we give a two-dimen-

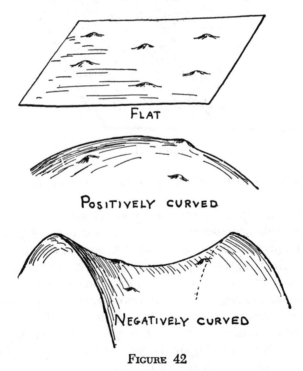

FLAT

POSITIVELY CURVED

NEGATIVELY CURVED

FIGURE 42

sional illustration of the flat space with "pimples," and two possible types of curved spaces. The so-called "positively curved" space corresponds to the surface of a sphere or any other closed geometrical figure, and bends "the same way" regardless of the direction in which one goes. The opposite type of "negatively curved" space bends up in one direction, but down in another, and resembles closely the surface of a western saddle. The difference between the two types of curvatures can be clearly realized if you cut out two pieces of leather, one from a football, another from the saddle, and try to straighten them out on a table. You will notice that neither can be straightened out without stretching or shrinking it, but, whereas the periphery of the football piece must be stretched, that of the saddle piece must be shrunk. The football piece has not enough material around the center to flatten it out; the saddle piece has too much, and it gets folded whenever we try to make it flat and smooth.

We can state the same point in still another way. Suppose we count the number of pimples located within one, two, three, etc. inches (counted along the surface) from a certain point. On the flat uncurved surface the number of pimples will increase as the square of the distances, that is, as 1, 4, 9, etc. On a spherical surface the number of pimples will increase more slowly than that, and on a "saddle" surface more rapidly. Thus the two-dimensional shadow scientists living *in* the surface, and thus having no way of looking at it from outside to notice its shape, will be still able to detect the curvature by counting the number of pimples that fall within the circles of different radii. It may be also noticed here that the difference between the positive and negative curvatures shows itself in the measurements of the angles in corresponding triangles. As we have seen in the previous section the sum of angles of a triangle drawn on the surface of a sphere is always *larger* than 180°. If you try to draw a triangle on the saddle surface you will find that the sum of its angles is always *less* than 180°.

The above results obtained specifically in regard to curved surfaces can be generalized in regard to curved three-dimensional spaces according to the table on page 112.

Type of space	Behavior at large distances	Angle-sum of a triangle	Volume of the sphere increases
Positively curved (sphere analogue)	Closes on itself	$>180°$	Slower than cube of the radius
Flat (plane analogue)	Extends into infinity	$=180°$	as the cube of the radius
Negatively curved (saddle analogue)	Extends into infinity	$<180°$	faster than cube of the radius

The table can be used in searching for a practical answer to the question as to whether the space in which we live is finite or infinite—a question that is discussed in Chapter X, which considers the size of the universe.

PART III

Microcosmos

CHAPTER VI

Descending Staircase

1. *THE GREEK IDEA*

IN ANALYZING the properties of material bodies, it is a good plan to start with some familiar object of "normal size," and to proceed step by step into its interior structure where the ultimate sources of all material properties lie hidden from the human eye. So let's begin our discussion with a bowl of clam chowder served on your dinner table. We have chosen the clam chowder not so much because it is tasty and nutritious, as because it represents a nice example of what is known as *heterogeneous* material. Even without the aid of a microscope you can see that it represents a mixture of a large number of different ingredients: the small slices of clams, the pieces of onions, tomatoes, and celery, the finely granulated potatoes, the particles of pepper, and little globules of fat, all mixed together in the salty watery solution.

Most of the substances, especially the organic ones, that we encounter in ordinary life are heterogeneous even though in many cases we need a microscope to help us recognize that fact. Even a small degree of magnification will show you that milk, for example, is a thin emulsion formed by small droplets of butter suspended in a uniform whitish liquid.

Ordinary garden soil is a fine mixture of microscopic particles of limestone, kaolin, quartz, iron oxide, and other minerals and salts, together with various organic substances derived from decayed plant and animal matter. And if we polish the surface of an ordinary granite rock, we shall see at once that this stone is formed by small crystals of three different substances (quartz, feldspar, and mica), strongly cemented together into one solid body.

In our study of the intrinsic structure of matter, the constitution of heterogeneous materials represents only the first step, or rather the upper landing of our descending staircase, and in each such case we can proceed directly to investigation of the individual *homogeneous* ingredients forming the mixture. Of really homogeneous substances such as a piece of copper wire, a glass of water, or the air filling the room (considered apart from the suspended dust, of course), no microscopic investigation will show any trace of different constituent parts, and the material will appear continuous throughout. True, in the case of copper wire, as in the case of practically every solid body (except those composed of glassy materials which do not crystallize), strong magnification always reveals a so-called microcrystalline structure. But the separate crystals we see in homogeneous materials are all of the same nature—crystals of copper in copper wire, crystals of aluminum in aluminum pans, etc.—exactly as in a tightly compressed handful of table salt we shall find only crystals of sodium chloride. By using a special technique (slow crystallization) we may increase the size of crystals of salt, copper, aluminum, or any other homogeneous substance to any desired extent, and a piece of such "monocrystalline" substance will be just as homogeneous throughout as water or glass.

Would we be justified by these observations, made both with the naked eye and with the best available microscopes, in assuming that the substances we call homogeneous will look the same no matter what degree of magnification we use? In other words, can we believe that no matter how small an amount of copper, salt, or water we have, they will always have the same properties as the larger samples, and always can be further subdivided into still smaller fragments?

The man who first formulated this question, and tried to give the answer to it, was the Greek philosopher Democritus who lived in Athens about twenty-three centuries ago. His answer to the question was in the negative; he was more inclined to believe that no matter how homogeneous a given substance may look, it must be considered to be formed by a large number (how large, he did not know) of separate very small particles (how small, he did not know either) which he called "atoms" or "indivisibles." These atoms, or indivisibles, differed in quantity in various substances, but their differences in quality were only apparent and not real. Fire atoms and water atoms were the same in fact, differing only in appearance. Indeed all materials were composed of the same eternal atoms.

Differing somewhat from this view, a contemporary of Democritus, named Empedocles, believed that there were several different kinds of atoms, which, mixed in different proportions, formed all the multitude of different known substances.

Reasoning on the basis of the rudimentary facts of chemistry known at that time Empedocles recognized four different types of atoms corresponding to four different alleged elementary substances: stone-stuff, water-stuff, air-stuff and fire-stuff.

According to these views, the soil for example was a combination of stone-stuff and water-stuff closely mixed atom by atom: the better the mixture, the better the soil. A plant growing from the soil combined stone and water atoms with the fire atoms coming from the rays of the sun to form composite molecules of wood-stuff. The burning of dry wood, from which the water element was gone, was viewed as decomposition or breaking up of wood molecules into the original fire atoms, which escape in the flame, and the stone atoms, which remains as the ashes.

We know now that this explanation of plant growth and wood burning, which would have looked quite logical at this early epoch in the infancy of science, is actually wrong. We know that plants take the largest part of the material used in the growth of their bodies not from the soil, as the ancients thought and as you may also think if nobody has yet told you otherwise, but from the air. Soil itself, apart from giving support to the growing plant and acting as a reservoir to hold the water which it needs, contributes only a very small proportion of certain salts necessary for plant growth, and one can grow a very big plant of corn from the amount of soil contained in a little thimble.

The truth is that the atmospheric air, which is a mixture of nitrogen and oxygen (and not a simple element as the ancients thought) contains also a certain amount of carbon dioxide, the molecules of which are formed by atoms of oxygen and atoms of carbon. Under the action of sunlight, the green leaves of the plant absorb atmospheric carbon dioxide, which reacts with the water supplied through the roots to form various organic materials from which the body of the plant is built. The oxygen is partially returned to the atmosphere, the process being responsible for the fact that "plants in the room refresh the air."

When wood burns, the molecules of wood-stuff unite again with the oxygen from the air, turning again into carbon dioxide and water vapor, which escape in the hot flame.

As to the "fire atoms" that the ancients believed entered into the material structure of the plant, they do not exist. The sunlight furnishes only the *energy* that is needed to break up molecules of carbon dioxide, thus making this atmospheric food digestible by the growing plant; and, since fire atoms do not exist, obviously their "escape" is not the explanation of fire; flame is simply the massed stream of gases heated up and made visible by the energy liberated in the process.

Let us now take another example illustrating similar differences between ancient and modern views on chemical transformations. You know of course that different metals are obtained from corresponding ores by subjecting them to very high temperatures in blast furnaces. At first sight most ores do not seem to differ much from ordinary rocks, so it is not surprising that ancient scientists believed that ores were made from the same stone-stuff as any other rock. Yet when they put a piece of iron ore into a hot fire they found that there came from it something quite different from ordinary rock—a strong shining substance of which good knives and spearheads could be made. The simplest way to explain this phenomenon was to say that the metal was formed by a union of stone and fire—or in other words, that the molecules of metal combined in their substance stone and fire atoms.

Having thus accounted for metals in general, they explained the different qualities of different metals such as iron, copper, and gold by saying that different proportions of stone and fire atoms went into their formation. Wasn't it obvious that shining gold contained more fire than darkish, dull iron?

But if this were so, why not add more fire to the iron, or still better to the copper, and thus turn them into the precious gold? Reasoning thus, practical-minded alchemists of the Middle Ages spent much of their lives over their smoky hearths trying to make "synthetic gold" from cheaper metals.

From their point of view their work was just as reasonable as that of a modern chemist who is developing a method for producing synthetic rubber; the fallacy of their theory and practice lay in their belief that gold and other metals were composite, rather than elementary substances. But how could one know which substance was elementary and which composite without trying? Had it not been for the futile attempts of these early chemists to turn iron or copper into gold or silver we might never have learned that metals are elementary chemical substances and that metal-bearing ores are composites formed by a combination of the atoms of metals and oxygen (metal oxides as the modern chemist says).

The transformation of iron ore into metallic iron under the sizzling heat of a blast furnace is not due to a union of atoms (stone atoms and fire atoms) as ancient alchemists thought, but quite the contrary, a result of a separation of atoms, that is, the removal of oxygen atoms from the composite molecules of iron oxide. The rust that appears on the surfaces of iron objects exposed to dampness is not composed of stone atoms left behind when fire atoms escape during the decomposition of iron-stuff, but to the formation of composite molecules of iron oxide resulting from the union of iron atoms and oxygen atoms from the air or water.[1]

From the above discussion it is apparent that ancient scientists' conceptions of the inner structure of matter and the nature of chemical transformation were basically correct; their error lay in

[1] Thus whereas an alchemist would express the processing of iron ore by the formula:

$$(\underbrace{\text{stone atom}) + (\text{fire atom}}_{\text{ore}}) \rightarrow (\text{iron molecule})$$

and the rusting of iron by:

$$(\text{iron molecule}) \rightarrow (\underbrace{\text{stone atoms}) + (\text{fire atoms}}_{\text{rust}})$$

we write for the same processes:

$$(\underset{\text{iron ore}}{\text{iron oxide molecule}}) \rightarrow (\text{iron atoms}) + (\text{oxygen atoms})$$

and

$$(\text{iron atoms}) + (\text{oxygen atoms}) \rightarrow (\underbrace{\text{iron oxide molecule}}_{\text{rust}})$$

a misconception of what constituted basic elements. In fact none of the four kinds of matter that Empedocles listed as elementary is in reality elementary; air is a mixture of several different gases, water molecules are formed from hydrogen and oxygen atoms, rocks have a very complex composition involving a great many different elements, and finally fire atoms do not exist at all.[2]

Actually there exist in nature not four but ninety-two different chemical elements, that is ninety-two different kinds of atoms. Some of these 92 chemical elements such as oxygen, carbon, iron, and silicon (the principal ingredients of most rocks) are rather abundant on the Earth and are familiar to everybody; others are very rare. You have probably never even heard of such elements as praseodymium, dysprosium or lanthanum. In addition to natural elements modern science has succeeded in making artificially several entirely new chemical elements, which we shall consider a little later in this book, one of which, known as *plutonium,* is destined to play an important role in the release of atomic energy for both warlike and peaceful uses. Combining among themselves in various proportions, the atoms of ninety-two basic elements form the unlimited number of various complex chemical materials such as water and butter, oil and soil, stones and bones, tea and TNT, and many other stuffs like triphenylpiriliumchloride and methylisopropylcyclohexane—terms which a good chemist must know by heart, but which most persons wouldn't even try to pronounce in one breath. And volumes upon volumes of chemical handbooks are being written to summarize the properties, the methods of preparation, and so forth of all this unlimited display of atomic combinations.

2. HOW LARGE ARE THE ATOMS?

When Democritus and Empedocles spoke of atoms they were essentially basing their arguments on vague philosophical ideas concerning the impossibility of imagining a process in which matter could be divided into smaller and smaller pieces without ever arriving at an indivisible unit.

[2] As we shall see later in this chapter, the idea of fire atoms was partially regenerated in the theory of light quanta.

When a modern chemist speaks of atoms, he means something much more definite, since precise knowledge of elementary atoms and their combination in complex molecules is absolutely necessary to the understanding of a fundamental law of chemistry according to which different chemical elements unite only in well-defined proportions by weight, the proportions that must apparently reflect the relative weights of the separate atoms of these substances. Thus the chemist concludes for example that the atoms of oxygen, aluminum, and iron must be respectively sixteen, twenty-seven, and fifty-six times heavier than the atoms of hydrogen. But, whereas the *relative atomic weights* of different elements represent the most important piece of basic chemical information, the actual weights of atoms, expressed in grams, are absolutely immaterial in chemical work, and knowledge of these exact weights would not in any way affect other chemical facts or application of the laws, and the methods of chemistry.

However, when a physicist considers atoms, his first question is bound to be: "What is the actual size of atoms in centimeters, how much do they weigh in grams, and how many individual atoms or molecules are there in a given amount of material? Is there any way to observe, to count, and to handle atoms and molecules individually one by one?"

There are many different ways of estimating the size of atoms and molecules, and the simplest of them is *so* simple that Democritus and Empedocles working without modern laboratory equipment could probably have used it had they happened to think about it. If the smallest unit in the composition of any material object, say a piece of copper wire, is an atom, it must obviously be impossible to make of this material a sheet thinner than the diameter of one such atom. Thus we can try to stretch our copper wire until it finally represents a chain of single atoms, or we can hammer it into a thin copper-leaf one atomic diameter thick. With copper wire, or any other solid material this task is next to impossible because the material will inevitably break before the desired minimum thickness is achieved. But liquid materials such as a thin layer of oil on the surface of water may easily be spread in a layer composed of a single blanket, so to speak, of its molecules, a mere film in which "individual" molecules join one

another horizontally, but in which none are piled on others vertically. With care and patience the reader can make this experiment himself, thus measuring by simple means the size of oil
molecules.

Take a shallow long vessel (Figure 43), place it on a table
or floor so that it is absolutely level, fill it with water to the rim,
and put across it a wire that will touch the surface of the water. If
you now drop a small droplet of some pure oil on one side of the
wire, the oil will spread all over that part of the surface of the
water that is on the side of the wire on which you have dropped
the oil. If you now move your wire along the rim of the vessel,
away from the oil, the layer of oil will spread, following the wire

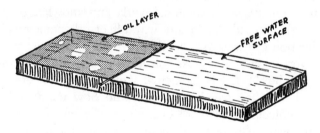

FIGURE 43

A thin oil layer on the water surface breaks up when stretched too
much.

and becoming thinner and thinner and its thickness must ultimately become equal to the diameter of a single oil molecule.
Any further motion of the wire after this thinness is achieved
will result in the breaking up of the continuous oil surface and
the formation of water holes. Knowing the amount of oil you put
on the water, and the maximum area over which it can spread
without breaking up, you can calculate easily the diameter of a
single molecule.

While performing this experiment, you may observe another
interesting phenomenon. When you drop some oil on the free
water surface you will notice first the familiar rainbow coloring
of the oil surface, just as you have probably seen it many times
on water in harbors frequented by ships. This coloring is due

to the well-known phenomenon of the interference of light rays reflected from the upper and lower boundaries of the oil layer, and the difference of color in different places is due to the fact that the oil layer spreading from the spot where the drop was placed has different thicknesses at different spots. If you wait a little until the layer becomes uniform, the entire oil surface will attain a uniform coloring. As the oil layer becomes thinner the coloring will gradually change from red to yellow, from yellow to green, from green to blue, and from blue to violet in conformance with the decreasing wavelength of light. If we continue to extend the area of the oil surface the coloring will entirely disappear. This does not mean that the oil layer is not there, but simply that its thickness has become less than the shortest visible wavelength, and the coloring goes out of the range of our vision. But you will still be able to distinguish the oily surface from the clear surface of water, since the two beams of light reflected from the upper and lower surfaces of a very thin layer will interfere in a way that leads to the reduction of total intensity. Thus when the coloring disappears the oily surface will differ from the pure surface by appearing somewhat more "dull" in the reflected light.

In actually performing this experiment, you will find that 1 cu mm of oil can cover about 1 sq m of water surface, but that any further attempt to stretch the oil film will lead to the formation of clear-water areas.[3]

3. *MOLECULAR BEAMS*

Another interesting method of demonstrating the molecular structure of matter can be found in the study of the outflow of

---

[3] How thin, then, is our oil layer just before it breaks up? In order to follow the calculations involved, envisage the droplet containing 1 cu mm of oil as an actual cube, each side of which is 1 sq mm. In order to stretch our original 1 cu mm of oil over the area of 1 sq m the 1-mm-square surface of the oil cube that is in contact with the water surface must be increased by a factor of a thousand (from 1 sq mm to 1 sq m). Consequently the vertical dimension of the original cube must be reduced by a factor $1000 \times 1000 = 10^6$ in order to keep the total volume constant. This gives us for the limiting thickness of the layer, and consequently for the actual size of oil molecules, the value of about $0.1 \text{ cm} \times 10^{-6} = 10^{-7}$ cm. Since an oil molecule consists of several atoms, the size of atoms is somewhat smaller.

gases and vapors through small openings into the surrounding empty space.

Suppose we have a large well-evacuated glass bulb (Figure 44) inside which is placed a small electric furnace consisting of a clay cylinder with a little hole in its wall, the cylinder surrounded by an electric resistance wire to furnish heat. If we place in the furnace a piece of some low-melting metal like sodium or potassium, the interior of the cylinder becomes filled with a metal vapor, which will leak out into surrounding space through the

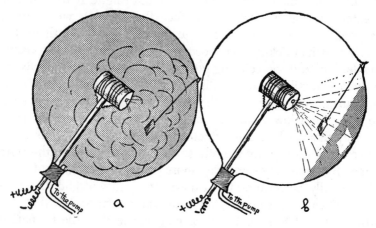

FIGURE 44

little hole in the cylinder's wall. Coming into contact with the cold walls of the glass bulb, the vapor will stick to them, and the thin mirrorlike deposit formed on various parts of the wall will clearly show us the way the material travels after escaping from the furnace.

Further, we shall see that the distribution of the film on the glass wall will differ, at different temperatures of the furnace. When the furnace is very hot, so that the density of the metal vapor inside it is rather high, the phenomenon will look familiar to anyone who has watched the steam escaping from a teakettle or a steam engine. Coming out through the opening, the vapor will expand in all directions (Figure 44a), filling up the entire volume of the bulb, and forming a more or less uniform deposit over the entire outer surface.

At lower temperatures, however, when the density of vapor inside the furnace is low, the phenomenon proceeds in an entirely different way. Instead of spreading out in all directions, the substance escaping through the hole seems to move along a straight line, and most of it is deposited on the glass wall facing the opening in the furnace. This fact can be particularly emphasized by placing some small object in front of the opening (Figure 44b). No deposit will be formed on the wall behind the object, and this region clear of deposit will have the exact shape of the geometrical shadow of the obscuring object.

The difference in behavior between gases escaping at high and low densities can be easily understood if one remembers that the vapor is formed by a very large number of separate molecules rushing through space in all directions and continuously colliding with one another. When the density of the vapor is high the stream of gas coming out through the opening can be compared with a frenzied crowd rushing through the exit doors of a burning theater. Having passed the doors, the people still bump into one another as they scatter in all directions on the street. When the density is low, on the other hand, it is as though only one person were passing through the door at one time and therefore proceeded straight ahead without interference.

The stream of matter of low vapor density coming through the furnace opening is known as a "molecular beam" and is formed by a large number of separate molecules flying through space side by side. Such molecular beams are very useful in a study of the individual properties of molecules. For example, one can use it to measure the velocity of thermal motion.

A device for the study of the velocity of such molecular beams was first built by Otto Stern, and is practically identical with that used by Fizeau for measuring the speed of light (see Figure 31). It consists of two cogwheels mounted on a common axis, and so arranged as to allow a molecular beam to pass through them only when the angular velocity of rotation is just right (Figure 45). By intercepting with a diaphragm a thin molecular beam from such an apparatus, Stern was able to demonstrate that the velocity of molecular motion is generally very high (1.5 km per second for sodium atoms at 200° C), and that it increases as the temperature of the gas rises. This furnishes a direct proof of the kinetic theory of heat, according to which the increase of the heat of a body is

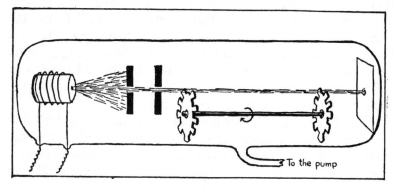

FIGURE 45

merely an increase in the irregular thermal motion of its molecules.

4. *ATOMIC PHOTOGRAPHY*

Although the above examples can hardly leave any doubt of the correctness of the atomic hypothesis, it is still true that "seeing is believing"; so that the most convincing evidence of the existence of atoms and molecules would lie in the tiny units themselves as seen by human eyes. Such a visual demonstration has been achieved only comparatively recently by the British physicist W. L. Bragg, who developed a method of obtaining photographs of separate atoms and molecules in various crystalline bodies.

One must not think, however, that photographing atoms is an easy job, since in taking pictures of such small objects one has to take into account the fact that the picture will be hopelessly blurred unless the wavelength of illuminating light is smaller than the size of the object to be photographed. There is no way to paint a Persian miniature with a housepainting brush! Biologists, who work with tiny micro-organisms, know this difficulty very well since the size of bacteria (about 0.0001 cm) is comparable to the wavelength of visible light. In order to improve the sharpness of the image they take their microphotographs of bacteria in ultraviolet light, thus obtaining somewhat better results than they might otherwise. But the size of molecules and their distances apart in a crystal lattice is so small (0.00000001 cm) that neither visible nor ultraviolet light is of any use when they are asked to sit for their portraits. In order to see the molecules separately we must necessarily use radiation with a wavelength thousands of times shorter than that of visible light—or, in other words, we have to use the radiation known as X rays.

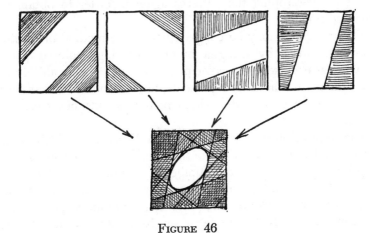

FIGURE 46

But here we encounter a seemingly insurmountable difficulty: X rays will pass through any substance practically without refraction, so that neither a lens nor a microscope will function when used with X rays. This property, together with the great penetrating power of X rays, is, of course, very useful in medical science, since refraction of rays while passing through the human body would completely blur all X-ray pictures. But the same property seems to exclude any possibility of getting any enlarged picture by means of X rays!

At first sight the situation seems hopeless, but W. L. Bragg found a very ingenious way out of the difficulty. He based his considerations on the mathematical theory of the microscope, developed by Abbé, according to which any microscopic image may be considered as the overlapping of a large number of separate patterns, each pattern being represented by parallel dark bands running at a certain angle through the field. A simple example illustrating the above statement can be seen in Figure 46, which shows how a picture of a luminous elliptical area in the center of the dark field can be obtained by the overlapping of four separate band systems.

According to Abbé's theory the functioning of a microscope consists in (1) breaking the original picture into a large number of separate band patterns, (2) enlarging each individual pattern, and (3) overlapping the patterns again in order to obtain the enlarged image.

The procedure may be compared with the method of printing colored pictures by using a number of single-colored plates. Looking at each separate color print, you may not be able to tell what the picture actually represents, but as soon as they all overlap in a proper way, the whole picture stands out clear and sharp.

The impossibility of constructing an X-ray lens which would perform all these operations automatically forces us to proceed step by step: taking a large number of separate X-ray band patterns of the crystal from all different angles, and then overlapping them in a proper way on one piece of photographic paper. Thus we can do exactly the same as an X-ray lens would do, but whereas the lens would do it almost instantaneously it would require a skillful experimenter many hours. This is why using Bragg's method we can make a picture of crystals, in which the molecules stay in their places, but cannot photograph them in liquids or gases, where they rush around in a wild dance.

Although the pictures made by Bragg's method are not actually obtained by a single click of the camera, they are as good and correct as any composite picture could be. Nobody would object to a photograph of a cathedral composed of several separate pictures, if for technical reasons one could not photograph the entire structure on one plate!

In Plate I we see a similar X-ray picture of a molecule of hexamethylbenzene, for which chemists write the formula:

$$
\begin{array}{ccccc}
 & H & & H & \\
 & | & & | & \\
 H-C-H & H-C-H & & \\
 & | & & | & \\
 & C\text{------}C & & H \\
 H & / & \backslash & | \\
 | & & & \\
 H-C-C & & C-C-H \\
 | & \backslash & / & H \\
 H & C\text{------}C & & H \\
 & | & & | & \\
 H-C-H & H-C-H & & \\
 & | & & | & \\
 & H & & H &
\end{array}
$$

The ring formed by six carbon atoms and another six carbon atoms attached to it stands out clearly in the picture, whereas the impressions of lighter hydrogen atoms are barely visible.

Even doubting Thomas, after seeing with his own eyes such photographs as these, would agree that the existence of molecules and atoms had been proved.

5. DISSECTING THE ATOM

When Democritus gave the atom its name, which in Greek means "indivisible," he meant that these particles represent the ultimate possible limit to which the breaking up of matter into its component parts could be carried, atoms, in other words, being the smallest and simplest structural parts of which all material bodies are composed. When thousands of years later the original philosophical idea of "an atom" was incorporated into the exact science of matter, and given flesh and blood on the basis of extensive empirical evidence, the belief in atomic indivisibility went along with it, and different properties of the atoms of various elements were hypothetically attributed to their different geometrical shapes. Thus, for example, the atoms of hydrogen were considered as being nearly spherical, whereas the atoms of sodium and potassium were believed to have the shapes of elongated ellipsoids.

The atoms of oxygen on the other hand were thought to have the shape of a doughnut with an almost completely closed central hole, so that a water molecule (H_2O) could be formed by placing two spherical hydrogen atoms into the holes on either

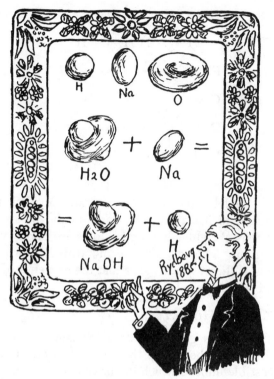

FIGURE 47

side of the oxygen doughnut (Figure 47). The substitution of sodium or potassium for hydrogen in a water molecule was then explained by the statement that the elongated atoms of sodium and potassium could be fitted better into the hole of the oxygen doughnut than the spherical atoms of hydrogen.

According to these views, the differences in optical spectra emitted by different elements was ascribed to the differences of the vibration frequencies of the differently shaped atoms. Reasoning thus, the physicists have tried unsuccessfully to draw conclusions about the shapes of different atoms composing light-emitting elements from the observed frequencies of the light which they emit, in the very same way as we explain in acoustics the differences of the sounds produced by a violin, a church bell, and a saxophone.

However, none of these attempts to explain the chemical and physical properties of various atoms exclusively on the basis of their geometrical shapes, led to any significant progress, and the first real step forward in an understanding of atomic properties was taken when it was recognized that atoms are not simple elementary bodies of various geometrical shapes, but on the con-

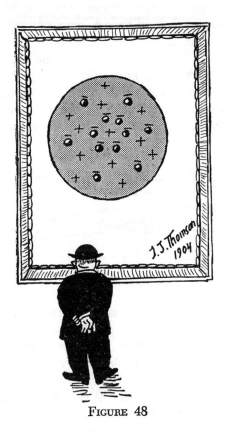

FIGURE 48

trary rather complex mechanisms with a large number of independent moving parts.

The honor of making the first incision in the complicated operation of dissecting the delicate body of the atom belongs to the famous British physicist J. J. Thomson, who was able to show that the atoms of various chemical elements consist of positively and negatively charged parts, held together by the forces of electrical attraction. Thomson conceived an atom as a more or less uniformly distributed positive electric charge with a large number of negatively charged particles floating in its interior (Figure 48). The combined electric charge of negative particles, or of *electrons* as he called them, equals the total positive charge, so that the atom on the whole is electrically neutral. Since, however, electrons were assumed to be bound comparatively loosely to the body of the atom, one or several of them could be removed leaving behind a positively charged atomic residue known as *positive ions*. On the other hand, the atoms that manage to get into their structure several extra electrons from the outside have an excess of negative charge and are known as *negative ions*. The process of communicating to an atom a positive or negative excess of

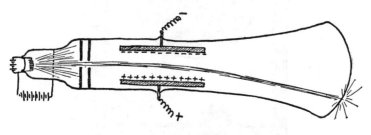

FIGURE 49

electricity is known as the process of *ionization*. Thomson based this view on the classical work of Michael Faraday, who had proved that whenever the atom carries an electrical charge it is always a multiple of a certain elementary amount of electricity numerically equal to 5.77×10^{-10} electrostatic unit. But Thomson went much farther than Faraday by ascribing to these electric charges the nature of individual particles, by developing the methods of their extraction from atomic bodies, and by studying the beams of free electrons flying at high speed through space.

A particularly important result of Thomson's studies of free electron beams was the estimate of their mass. Into the space between the two plates of a charged condenser (Figure 49), he sent a beam of electrons extracted by a strong electric field from some such material as hot electric wires. Being charged with negative electricity, or to put it more correctly, being the free negative charges themselves, the electrons of the beam were attracted to the positive electrode and repelled by the negative one.

The resulting deflection of the beam could easily be observed by allowing it to fall on a fluorescent screen placed behind the condenser. Knowing the charge of an electron, and its deflection in a given electric field, it was possible to estimate its mass, which turned out to be very small indeed. In fact, Thomson found that the mass of one electron is 1840 times smaller than the mass of a hydrogen atom, thus indicating that the main portion of atomic mass is contained in its positively charged parts.

Being quite right in his views about the swarm of negative electrons moving inside the atom, Thomson was, however, very far from the truth concerning the uniform distribution of a positive charge through the body of the atom. It was shown by Rutherford in 1911 that the positive charge of the atom as well as the largest part of its mass is concentrated in an extremely small *nucleus* located in the very center of the atom. He came to this conclusion as the result of his famous experiments on the scattering of the so-called "alpha (α) particles" in their passage through matter. These α-particles are tiny high-speed projectiles emitted by the spontaneously breaking up of atoms of certain heavy unstable elements (like uranium or radium), and, since their mass was proved to be comparable with the mass of atoms and their charge is positive, they must be considered as the fragments of the original positive body of the atom. As an α-particle passes through the atoms of the target material, it is influenced by the forces of attraction toward atomic electrons and the forces of repulsion from the positive parts of the atom. Since, however, the electrons are so exceedingly light, they are no more able to influence the motion of the incident α-particle, than a swarm of mosquitoes can influence the run of a scared elephant. On the other hand the repulsion between the massive positive parts of the atom and the positive charge of incident α-particles must be able to deflect the latter from their ordinary trajectory and to scatter them in all directions, provided they pass sufficiently close by one another.

In studying the scattering of a beam of α-particles passing through a thin filament of aluminum, Rutherford came to the surprising conclusion that in order to explain the observed results one must assume that the distance between the incident α-particles and the positive charge of the atom becomes smaller than one thousandth of the atomic diameter. This can be, of course, only possible if *both the incident alpha particles and the posi-*

FIGURE 50

tively charged part of the atom are thousands of times smaller than the atom itself. Thus Rutherford's discovery shrank the originally widespread positive charge of Thomson's atomic model into a tiny *atomic nucleus* in the very center of the atom, leaving the swarm of negative electrons on the outside, so that instead of being similar to a watermelon with electrons playing the role of seeds, the picture of an atom began to look more like a miniature solar system with an atomic nucleus for the sun, and electrons for planets (Figure 50).

The analogy with the planetary system can be further strengthened by these facts: the atomic nucleus contains 99.97 per cent of the total atomic mass as compared with 99.87 per cent of the solar system concentrated in the sun, and the distances between the planetary electrons exceed their diameters by about the same factor (several thousand times) which we find when comparing interplanetary distances with the diameters of the planets.

The more important analogy lies, however, in the fact that the electric attraction-forces between the atomic nucleus and the electrons obey the same mathematical law of inverse square[4] as the gravity forces acting between the sun and the planets. This makes the electrons describe the circular and elliptic trajectories around the nucleus, similar to those along which the planets and comets move in the solar system.

According to the foregoing views concerning the internal structure of the atom, the difference between the atoms of various chemical elements must be ascribed to the different number of electrons rotating around the nucleus. Since the atom as a whole is electrically neutral, the number of electrons rotating around its nucleus must be determined by the number of elementary positive charges carried by the nucleus itself, which number, in its turn, can be estimated directly from the observed scattering of α-particles deflected from their tracks by electric interaction of the nuclei. It was found by Rutherford that *in the natural sequence of chemical elements arranged in the order of increasing weights there is a consistent increase of one atomic electron in each element in the sequence.* Thus an atom of hydrogen has 1 electron; an atom of helium 2; lithium 3; beryllium 4; and so on up to the heaviest natural element, uranium, which has altogether 92 electrons.[5]

This numerical designation of an atom is usually known as the *atomic number* of the element in question, and coincides with its

[4] That is, the forces are inversely proportionate to the square of the distance between two bodies.

[5] Now that we have learned the art of alchemy (see later) we can build artificially even more complex atoms. Thus the artificial element *plutonium* used in atomic bombs has ninety-four electrons.

positional number, which indicates its position in a classification
in which the elements have been arranged by chemists according
to their chemical properties.

Thus all the physical and chemical properties of any given

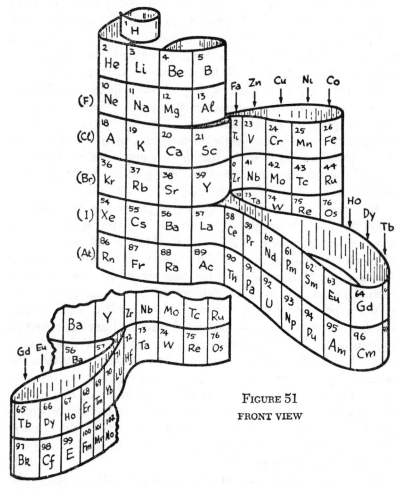

FIGURE 51
FRONT VIEW

element can be characterized simply by the one figure giving the
number of electrons rotating around the central nucleus.

Toward the end of the last century the Russian chemist
D. Mendeleev noticed a remarkable periodicity in the chemical
properties of elements arranged in a natural sequence. He found

that the properties of the elements begin to repeat themselves after a definite number of steps. This periodicity is represented graphically in Figure 51, in which the symbols of all presently known elements are represented along a spiral band on the surface

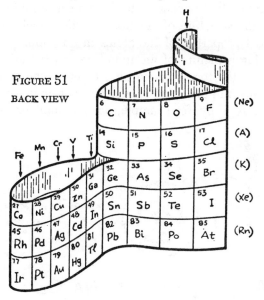

The periodic system of elements arranged on a wound ribbon, showing the periods 2, 8, and 18. The lower diagram on the opposite page represents the other side of the loop of elements (rare earths and actinides) which fall out from regular periodicity.

of a cylinder in such a way that the elements with similar properties are located in columns. We see that the first group contains only 2 elements: hydrogen and helium; then we have two groups of 8 elements each; and finally the properties repeat themselves after every 18 elements. If we remember that each step along the sequence of elements corresponds to one additional electron in the atom, we must inevitably conclude that the observed periodicity of chemical properties must be due to the recurrent formation of certain stable configurations of atomic electrons, or "electronic shells." The first completed shell must consist of 2 electrons, the next two shells of 8 electrons each, and all the following shells of 18 each. We also notice from Figure 51 that in the sixth and seventh periods the strict periodicity of properties becomes a little confused and two groups of elements (the so-called rare earths and the actinides) must be placed on a band protruding from the regular cylindrical surface. This anomaly is due to the fact that we encounter here a certain internal reconstruction of the structure of electronic shells, which plays havoc with chemical properties of the atoms in question.

Now, having a picture of an atom, we can try to answer the question about the forces which bind together the atoms of different elements into the complex molecules of innumerable chemical compounds. Why, for example, do the atoms of sodium and

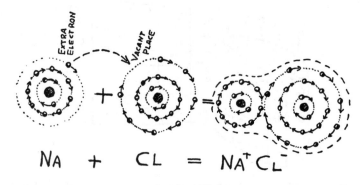

FIGURE 52

Schematic picture representing the union of sodium and chlorine atoms in a sodium chloride molecule.

chlorine stick together to form a molecule of table salt? We see from Figure 52 representing the shell structure of these two atoms that an atom of chlorine lacks one electron in order to complete the third shell, whereas an atom of sodium has one extra electron left after the completion of its second shell. Thus there must be the tendency for the extra electron from sodium to go over into chlorine to complete the unfinished shell. As the result of this transition of an electron, the sodium atom becomes positively charged (by losing a negative electron), whereas the atom of chlorine acquires a negative charge. Under the forces of electric attraction between them, the two charged atoms (or *ions* as they are called) will cling together forming a molecule of sodium chloride, or in plain words table salt. In the same way an atom of oxygen that lacks two electrons in its outer shell will "kidnap" from two hydrogen atoms their single electrons thus forming a molecule of water (H_2O). On the other hand, there will be no tendency to combine either between the atoms of oxygen and chlorine, or between those of hydrogen and sodium, since in the first case both have the desire to take and not to give, whereas in the second case neither wants to take.

The atoms with completed electronic shells, such as those of helium, argon, neon, and xenon, are completely self-satisfied and do not need to give or to take extra electrons; they prefer to remain gloriously lonely making the corresponding elements (so called "rare gases") chemically inert.

We conclude this section about atoms and their electronic shells by referring to the important role that atomic electrons play in the substances generally known under the collective name of "metals." The metallic substances differ from all other materials by the fact that the outer shells of their atoms are bound rather loosely, and often let one of their electrons go free. Thus the interior of a metal is filled up with a large number of unattached electrons that travel aimlessly around like a crowd of displaced persons. When a metal wire is subjected to electric force applied on its opposite ends, these free electrons rush in the direction of the force, thus forming what we call an electric current.

The presence of free electrons is also responsible for high heat conductivity—but we shall return to this subject in one of the following chapters.

6. *MICROMECHANICS AND THE UNCERTAINTY PRINCIPLE*

Since, as we have seen in the previous section, the atom with its system of electrons rotating around the central nucleus resembles closely the planetary system, it would be natural to expect that it should be subject to the same well-established astronomical laws that govern the motion of planets around the sun. In particular the similarity between the laws of electric and gravitational attraction—in both cases the attractive force is inversely proportionate to the square of the distance—would suggest that atomic electrons must move along elliptical orbits with the nucleus as the focus (Figure 53a).

However, all attempts to build a consistent picture of the motion of atomic electrons along the same pattern as that used in delineating the movements of our planetary system until recently led to an unexpected disaster of such magnitude that it looked for a while as though either the physicists or physics itself had become completely insane. The trouble arose essentially

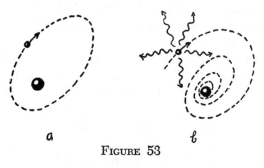

a   b

FIGURE 53

from the fact that, unlike the planets of the solar system, atomic electrons are electrically charged, and, as is true of any vibrating or rotating electrical charges, their circular motion around the nucleus must be expected to give rise to an intensive electro‑magnetic radiation. As a result of the loss of energy that is carried away by the radiation, it is logical to suppose that atomic electrons approach the nucleus along a spiral trajectory (Figure 53b), and finally fall on it when the kinetic energy of their orbital motion is completely exhausted. As to the time consumed in this process, it was a fairly simple matter to calculate from the known electric charge and the rotation frequency of atomic electrons that it should not take more than about one hundredth of a microsecond for the electrons to lose all their energy and to fall.

Thus, according to the best knowledge and belief of physicists until very recently, the planetary-like atomic structures should not be able to exist for more than a negligible part of a second and were doomed to an almost immediate collapse as soon as they were formed.

Yet, in spite of these grim predictions of physical theory, experiments showed that the atomic systems were very stable indeed, and that the atomic electrons kept right on happily circling their central nuclei without any loss of energy whatsoever, and without any tendency to collapse!

How could it be! Why should the application of old and well-established laws of mechanics to atomic electrons lead to conclusions so contradictory to observed facts?

To answer this question we have to turn to the most fundamental problem of science: the problem of the nature of science itself. What is "science," and what do we mean by the "scientific explanation" of the facts of nature?

To take a simple example, let us remember that many ancient peoples believed that the Earth was flat. One can hardly blame them for such a belief, because if you come out into an open field, or sail in a boat across the water, you will *see* for yourself that it is true; apart from occasional hills and mountains, the surface of the Earth *does look flat*. The mistake of the ancients was not in the statement that "the Earth is flat as far as one can see it from a given observation point," but in the *extrapolation* of this statement beyond the limits of actual observation. And, in fact, the observations that went far beyond the conventional limits, such as the study of the shape of the Earth's shadow on the Moon during the eclipse, or Magellan's famous expedition around the world, immediately proved the error of such extrapolation. We say now that the Earth *looks* flat only because what we can see represents a very small portion of the total surface of the globe. Similarly, as discussed in Chapter 5, the space of the universe may be curved and finite in size, in spite of the fact that it looks flat and apparently infinite from the point of view of limited observations.

But what has it all to do with the contradiction to which we came in studying the mechanical behavior of electrons forming the body of the atom? The answer is that, in these studies, we have assumed implicitly that the mechanism of an atom follows exactly the same laws as those that govern the motion of large celestial bodies, or, for that matter, the motion of bodies of "normal size" that we are accustomed to handle in everyday life, and thus may be described in the same terms. In fact, the familiar laws and concepts of mechanics were established empirically for material bodies comparable in size to human beings. The same laws were later used to explain the motion of much larger bodies, such as planets and stars, and the success of celestial mechanics, which permits us to calculate with the utmost precision various astronomical phenomena millions of years ahead and millions of years back in time, seems to leave no doubt of the validity of extrapolation of the customary mechanical laws in explaining the motion of large celestial masses.

But what assurance do we have that the same laws of mechanics, which explain the movements of giant celestial bodies, as well as those of artillery shells, clock pendulums, and toy spinning-tops, will also apply to the motion of electrons that are billions and billions times smaller and lighter than the tiniest mechanical device we ever had in our hands?

Of course, *there is no reason to suppose in advance that the laws of ordinary mechanics must fail in explaining the motion of the tiny constituent parts of the atom; but, on the other hand, one should also not be too greatly surprised if such a failure actually takes place.*

Thus, the paradoxical conclusions, which resulted from the attempt to determine the motion of atomic electrons in the same way as an astronomer explains the motion of planets in the solar system, must first of all be considered in the light of possible changes in the fundamental notions and laws of classical mechanics in its application to particles of such an exceedingly small size.

The basic concepts of classical mechanics are those of the *trajectory* described by a moving particle, and of the *velocity* with which a particle moves along its trajectory. The proposition that *any moving material particle occupies in any given moment a definite position in space, and that the consecutive positions of this particle form a continuous line known as the trajectory* was always considered as self-evident, and formed the fundamental basis for the description of the motion of any material body. The distance between two locations of a given object at different moments of time, divided by the corresponding time interval, led to the definition of *velocity*, and on these two concepts of location and velocity all classical mechanics was built. Until very recently it probably never occurred to any scientist that those most fundamental concepts used in the description of the phenomena of motion could be to any extent incorrect, and it was customary among the philosophers to consider them as given "a priori."

However, the complete fiasco that resulted from the attempts to apply the laws of classical mechanics to the description of motions within the tiny atomic systems showed that in this case something was basically wrong, and led to the ever-growing belief that this "wrongness" extended to the most fundamental ideas on which classical mechanics was based. The basic kinematic notions of the continuous trajectory of a moving object, and its well-defined velocity at any given moment of time seem to be *too rough* when applied to the tiny parts of internal atomic mechanisms. In short, the attempt of the extrapolation of the ideas of familiar classical mechanics into the region of exceedingly small masses proved conclusively that in doing so we have to change these ideas in a rather drastic way. But if the old notions of classical mechanics do not apply in the atomic world, they also cannot be *absolutely* correct in regard to the motion of larger material bodies. Thus we are led to the conclusion that *the principles underlying classical mechanics must be considered only as very good approximations to the "real thing,"* approximations that fail badly as soon as we try to apply them to systems more delicate than those for which they were originally intended.

The essentially new element that was brought into the science of matter by the study of the mechanical behavior of atomic systems, and by the formulations of the so-called quantum physics consisted in the discovery of the fact that *there is a certain lower limit for any possible interaction between two different material bodies,* which discovery plays havoc with the classical definition of the trajectory of a moving object. In fact, the statement that there exists such a thing as the mathematically exact trajectory of a moving object implies *the possibility of recording* this trajectory by means of some specially adapted physical apparatus. It must not, however, be forgotten that in recording the trajectory of any moving object we necessarily disturb the original motion; in fact if our moving object exercises some action on the measuring apparatus that records its successive positions in space, the apparatus acts back on the moving object according to the Newtonian law of the equality of action and reaction. If, as it was assumed in classical physics, the interaction between two material bodies (in this case between the moving object and the apparatus recording its position) can be made as small as desired, we may imagine an ideal apparatus so sensitive that it records the successive positions of the moving object with practically no disturbance of its motion.

FIGURE 54

Micromechanical pictures of the electronic motion in the atom.

The existence of the *lower limit of physical interactions* changes the situation in a rather radical way, since we cannot any more reduce the disturbance of motion caused by its recording to an arbitrarily small value. Thus *the disturbance of motion caused by its observation becomes an integral part of the motion itself*, and, instead of speaking about the infinitely thin mathematical line representing the trajectory, we are forced to substitute in its place a diffused band of a finite thickness. *The mathematically sharp trajectories of classical physics become broad diffused bands in the eyes of the new mechanics.*

The minimum amount of physical interaction, or the *quantum of action* as it is usually known, has however a very small numerical value and becomes of importance only when we study the motion of very tiny objects. Thus, for example, although it is true that the trajectory of a revolver bullet is not a mathematically sharp line, the "thickness" of this trajectory is many times smaller than the size of a single atom of the material forming the bullet, and thus can be assumed to be practically zero. However, turning to the lighter objects that are more easily influenced by the disturbances originating in the measurement of their motion, we find that the "thickness" of their trajectories becomes more and more essential. *In the case of atomic electrons rotating* around the central nucleus the thickness of the orbit becomes comparable to its diameter, so that, instead of representing their motion by a line as it was done in Figure 53, we are compelled to visualize it in the way shown in Figure 54. In such cases the motion of the particles cannot be described in the familiar terms of classical mechanics, and both its position and its velocity are subject to a certain indefiniteness (Heisenberg's uncertainty relations and Bohr's complimentarity principle).[6]

This startling development of the new physics, which throws into the wastepaper basket such familiar notions as trajectory of motion and the exact position and velocity of a moving particle, seems to leave us in thin air. If we are not permitted to use these formerly accepted basic principles in our study of atomic electrons, on what can we base our understanding of their motion? What is the mathematical formalism that must be substituted for the methods of classical mechanics in order to take care of the uncertainties of position, velocity, energy, and so on demanded by the facts of quantum physics?

The answer to these questions can be found by considering a similar situation that existed in the field of the classical theory of light. We know that most of the light phenomena observed in ordinary life can be interpreted on the basis of the assumption that light propagates along straight lines known as *light rays*. The shapes of shadows thrown by nontransparent objects, the formation of images in plane and curved mirrors, the functioning of lenses and various more complex optical systems can readily be explained on the basis of the elementary laws governing the reflection and refraction of light rays (Figure 55a, b, c).

But we also know that the methods of geometrical optics that attempt to demonstrate the classical theory of the propagation of light by means of light rays fail badly in cases where the geo-

[6] More detailed discussion of uncertainty relations can be found in the author's book *Mr. Tompkins in Wonderland* (The Macmillan Co., New York, 1940).

metrical dimensions of the openings used in the optical systems become comparable with the wavelength of light. The phenomena that take place in these cases, are known as *diffraction-phenomena* and fall entirely outside the scope of geometrical

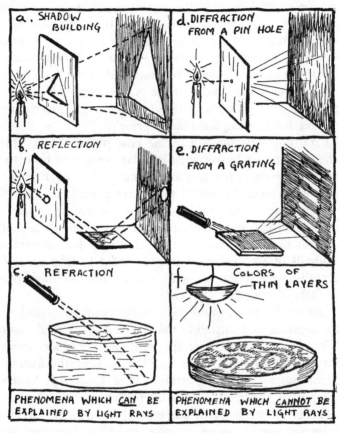

FIGURE 55

optics. Thus a light beam passing through a very small opening (of the order of 0.0001 cm) fails to be propagated along a straight line, and is instead scattered in a peculiar fanlike manner (Figure 55*d*). When a beam of light falls on a mirror covered with a large number of parallel narrow lines scratched on its surface ("diffracting grating"), it does not follow the familiar law of reflection, but is thrown in a number of different directions determined by the distances between the scratched lines and the wavelength of incident light (Figure 55e). It is also known that when light is reflected from thin layers of oil spread on the surface of water a peculiar system of light and dark fringes are created (Figure 55f).

In all these cases, the familiar notion of a "light ray" completely fails to describe the observed phenomena, and we must recognize instead a continuous distribution of light energy over the entire space occupied by the optical system.

It is easy to see that the failure of the notion of a light ray in application to the optical diffraction phenomena is very similar to the failure of the notion of a *mechanical trajectory* in the phenomena of quantum physics. Just as we cannot form in optics an infinitely thin light beam, the quantum principles of mechanics prevent us from speaking about the infinitely thin trajectories of moving particles. In both cases we have to give up all attempts to describe the phenomena by saying that something (light or particles) propagates along certain mathematical lines (optical rays or mechanical trajectories), and are forced to go over to the presentation of the observed phenomena by means of "something" which is spread continuously over the entire space. With respect to light this "something" is the intensity of light vibrations at various points; with respect to mechanics the "something" is the newly introduced notion of the uncertainty of location, the probability that a moving particle may be found at any given moment, not in a predetermined spot, but in any one of several possible locations. It is not possible any more to state *exactly* where a moving particle is at a given moment of time, though the limits within which such a statement can be made may be calculated by the formulas concerning "uncertainty relations." The relation that exists between the laws of wave optics concerned with the diffraction of light, and those of the new "micro-mechanics," or "wave mechanics" (developed by L. de Broglie and E. Schrödinger) concerned with the motion of mechanical particles, can be made very apparent by the experiments showing the similarity of these two classes of phenomena.

In Figure 56 we show the arrangement used by O. Stern in his study of atomic diffraction. A beam of sodium atoms, produced by a method described earlier in this chapter, is reflected from the surface of a crystal. The regular atomic layers forming the crystalline lattice act in this case as the diffraction grating for the incident beam of particles. The incident sodium atoms reflected from the surface of the crystal are collected into a series of little bottles placed at different angles, and the number of atoms collected in each bottle is carefully measured. The dotted line in Figure 56 represents the result. We see that instead of being reflected in one definite direction (as are ball bearings shot from a little toy gun onto a metal plate), sodium atoms are distributed

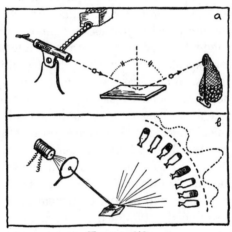

FIGURE 56

(a)   Phenomenon explainable by the notion of trajectory (reflection of ball bearings from a metal plate).
(b)   Phenomenon unexplainable by the notion of trajectory (reflection of sodium atoms from a crystal).

within well-defined angles forming a pattern very similar to that observed in ordinary X-ray diffraction.

Experiments of such kind cannot possibly be explained on the basis of classical mechanics, which describes the motion of separate atoms along definite trajectories, but are perfectly understandable from the point of view of the new micromechanics, which considers the motion of particles in the same ways as modern optics considers the propagation of light waves.

CHAPTER VII

Modern Alchemy

1. *ELEMENTARY PARTICLES*

HAVING learned that the atoms of various chemical elements represent rather complicated mechanical systems with a large number of electrons rotating around the central nucleus, we inevitably ask whether these atomic nuclei are the ultimate indivisible structural units of matter, or whether they in turn can be subdivided still farther into smaller and simpler parts. Would it be possible to reduce all the 92 different atomic types to perhaps a couple of really simple particles?

As early as the middle of the last century this desire for simplicity had driven an English chemist, William Prout, to a hypothesis according to which *the atoms of all different chemical elements have a common nature representing only various degrees of "concentration" of hydrogen atoms.* Prout based his hypothesis on the fact that the chemically determined atomic weights of various elements in respect to hydrogen are in most cases represented very closely by integer numbers. Thus according to Prout, the atoms of oxygen, which are 16 times heavier than those of hydrogen, must be considered as made up from 16 hydrogen atoms stuck together. The atoms of iodine with an atomic weight of 127 must be formed by an aggregate of 127 hydrogen atoms, etc.

However the findings of chemistry were at that time very unfavorable to the acceptance of this bold hypothesis. It was shown, by the exact measurements of atomic weights, that they could not be represented exactly by integer numbers but in most cases only by numbers very close to integers, and in a few cases by numbers that were not even close to integers. (The chemical atomic weight of chlorine, for instance, is 35.5.) These facts, which are seemingly in direct contradiction to Prout's hypothesis, discredited it and Prout died without ever learning how right he actually was.

It was not until the year 1919 that his hypothesis again asserted itself through the discovery of the British physicist F. W. Aston, who showed that ordinary chlorine represents a *mixture* of two different kinds of chlorine possessing identical chemical properties but having different integer atomic weights: 35 and 37. The noninteger number 35.5 obtained by chemists represents only the mean value for the mixture.[1]

The further study of various chemical elements revealed the striking fact that most of them are a mixture of several components identical in chemical properties but different in atomic weight. They received the name of *isotopes*, that is, substances occupying the same place in the periodic system of elements.[2] The fact that the masses of different isotopes are always the multiples of the mass of a hydrogen atom gave new life to Prout's forgotten hypothesis. Since, as we have seen in the previous section, the main mass of the atom is concentrated in its nucleus, Prout's hypothesis can be reformulated in modern language by saying that *the nuclei of different atomic species are composed of various numbers of elementary hydrogen nuclei, which, because of their role in the structure of matter, were given the special name of "protons."*

There is, however, one important correction to be made in the above statement. Consider for example the nucleus of an oxygen atom. Since oxygen is the eighth element in the natural sequence, its atom must contain 8 electrons and its nucleus must carry 8 positive elementary charges. But oxygen atoms are 16 times heavier than those of hydrogen. Thus, if we assume that an oxygen nucleus is formed from 8 protons, we would get a correct charge but a wrong mass (both 8); assuming 16 protons we get correct mass but wrong charge (both 16).

It is clear that the only way out of the difficulty lies in the assumption that *some of the protons forming complex atomic*

[1]Since the heavier chlorine is present in the amout of 25 per cent and the lighter in the amount of 75 per cent, the mean atomic weight must be: $0.25 \times 37 + 0.75 \times 35 = 35.5$, which is exactly what the earlier chemists had discovered.

[2] From the Greek ισος meaning equal and τοπος meaning place.

nuclei have lost their original positive charges and are electrically neutral.

The existence of such chargeless protons, or "neutrons" as they are called now, was suggested by Rutherford as early as 1920, but it was twelve years before they were found experimentally. It must be noted here that protons and neutrons should not be considered as two entirely different kinds of particles, but rather as two different electrical states of the same basic particle

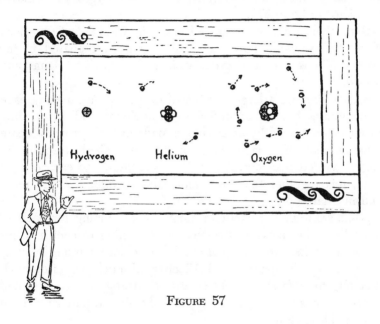

FIGURE 57

known now under the name of a "nucleon." In fact, it is known that protons can turn into neutrons by losing their positive charge, and neutrons can turn into protons by acquiring it.

The introduction of neutrons as the structural units of atomic nuclei removes the difficulty discussed on previous pages. In order to understand that the nucleus of an oxygen atom has 16 units of mass, but only 8 units of charge we must accept the fact that it is formed by 8 protons and 8 neutrons. The nucleus of iodine, with an atomic weight of 127 and the atomic number 53, consists of 53 protons and 74 neutrons, whereas the heavy nucleus of uranium (atomic weight: 238, atomic number: 92) is formed by 92 protons and 146 neutrons.[3]

Thus, almost a century after its origin, the bold hypothesis of Prout finally received the honorable recognition that it so well deserved, and we may now say that the infinite variety of known substances results from different combinations of only two kinds of fundamental particles: (1) *nucleons*, the basic particles of matter, which can be either neutral or can carry a positive electric charge; and (2) *electrons*, the free charges of negative electricity (Figure 57).

Here, then are a few recipes from "The Complete Cook Book of Matter" showing how each dish has been prepared in the Cosmic Kitchen from a larder well-stocked with nucleons and electrons:

WATER. Prepare a large number of oxygen atoms, making each by combining 8 neutral and 8 charged nucleons, and surrounding nucleus so obtained by an envelope made of 8 electrons. Prepare twice as many hydrogen atoms by attaching single electrons to single charged nucleons. Add 2 hydrogen atoms to each oxygen atom; mix together the water molecules so obtained and serve cold in a large glass.

TABLE SALT. Prepare sodium atoms by combining for each 12 neutral and 11 charged nucleons and attaching to each nucleus 11 electrons. Prepare an equal number of chlorine atoms by combining 18 or 20 neutral and 17 charged nucleons (isotopes), attaching to each nucleus 17 electrons. Arrange the sodium and chlorine atoms in a three-dimensional chessboard pattern to form regular salt crystals.

TNT. Prepare carbon atoms by combining 6 neutral and 6 charged nucleons with 6 electrons attached to the nucleus. Prepare nitrogen atoms from 7 neutral and 7 charged nucleons each with 7 electrons around the nucleus. Prepare oxygen and hydrogen atoms according to the prescription given above (see: WATER). Arrange 6 carbon atoms in a ring with a 7th carbon

[3] Looking through the table of atomic weights you will notice that at the beginning of the periodic system atomic weights are equal to twice the atomic number, meaning that these nuclei contain an equal number of protons and neutrons. For heavier elements atomic weights increase more rapidly indicating that there are more neutrons than protons.

atom outside the ring. Attach 3 pairs of oxygen atoms to 3 carbons of the ring placing in each case 1 nitrogen atom between oxygen and carbon. Attach 3 hydrogen atoms to the carbon outside the ring, and 1 hydrogen to each of the 2 vacant carbon places in the ring. Arrange the molecules so obtained in a regular pattern to form a large number of small crystals and press all these crystals together. Handle with care, since this structure is unstable and highly explosive.

Although, as we have just seen, *neutrons, protons,* and *negative electrons* represent the only necessary building units for the construction of any desired material substance, this list of basic particles seems still somewhat incomplete. In fact, if ordinary electrons represent free charges of negative electricity, why can't we also have free charges of positive electricity, that is, *positive electrons?*

Also, if a neutron, which apparently represents the basic unit of matter, can acquire a positive electric charge, thus becoming a proton, why cannot it also become negatively charged, forming a *negative proton?*

The answer is that positive electrons, which are quite similar to ordinary negative electrons except in the sign of their charge, actually do exist in nature. And there is also a certain possibility that negative protons exist, although experimental physics has not yet succeeded in detecting them.

The reason that positive electrons and negative protons (if any) are not as plentiful in our physical world as negative electrons and positive protons lies in the fact that these two groups of particles are, so to speak, antagonistic to each other. Everybody knows that two electrical charges, one of which is positive and the other negative, will cancel each other when they are put together. Thus, since the two kinds of electrons represent nothing else but free charges of positive and negative electricity, one should not expect them to coexist in the same region of space. In fact, as soon as a positive electron encounters a negative one, their electric charges will immediately cancel one another and the two electrons will cease to exist as individual particles. Such a process of mutual annihilation of two electrons results, however, in the birth of an intensive electromagnetic radiation (gamma [γ]

rays) escaping from the point of the encounter and carrying with it the original energy of the two vanished particles. According to a fundamental law of physics, energy can be neither created nor destroyed, and we are witnessing here only the transformation of the electrostatic energy of the free electric charges into the electrodynamic energy of the radiated wave. The phenomenon which results from the encounter of a positive and a negative electron is described by Prof. Born[4] as a "wild marriage" and

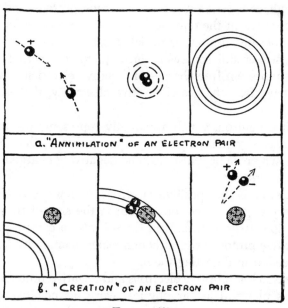

a. "Annihilation" of an electron pair

b. "Creation" of an electron pair

FIGURE 58

Schematic picture of the "annihilation" process of two electrons giving rise to an electromagnetic wave, and the "creation" of a pair by a wave passing close to an atomic nucleus.

by the more gloomy Prof. Brown[5] as the "mutual suicide" of the two electrons. Figure 58*a* is a graphic representation of the encounter.

The process of "annihilation" of two electrons with opposite charges, has its counterpart in the process of "pair formation," by which a positive and a negative electron are formed appar-

[4] M. Born, *Atomic Physics* (G. E. Stechert & Co., New York, 1935).
[5] T. B. Brown, *Modern Physics* (John Wiley & Sons, New York, 1940).

ently from nothing as a result of strong gamma radiation. We say "apparently" from nothing because actually each such newborn pair of electrons is formed at the cost of the energy supplied by γ-rays. In fact the amount of energy that must be given away by the radiation in order to form an electron pair is exactly the same as that which is set free in the process of annihilation. This process of *pair formation*, which takes place preferably when the incident radiation passes close to some atomic nucleus,[6] is represented schematically in Figure 58*b*. We have here an example of the formation of two opposite charges of electricity where there was originally no charge at all, a process that should not, however, be considered any more surprising than the familiar experiment in which an ebonite stick and a piece of wool cloth become charged with opposite electricities when rubbed against each other. Having a sufficient amount of energy, we can produce as many pairs of positive and negative electrons as we like, recognizing fully, however, the fact that the process of mutual annihilation will soon take them out of circulation again, paying back "in full" the amount of energy originally spent.

A very interesting example of such "mass production" of electron pairs is presented by the phenomenon of "cosmic-ray showers," which are produced in the terrestrial atmosphere by the streams of high-energy particles coming to us from interstellar space. Although the origin of these streams which crisscross in all directions the empty vastness of the universe is still one of the unsolved riddles of science,[7] we have a rather clear idea of what happens when the electrons moving with terrific

---

[6] Although in principle the formation of an electron pair can take place in a completely empty space, the process of pair formation is considerably helped by the presence of the electric field surrounding atomic nuclei.

[7] The most trivial, but probably also the most plausible, explanation of the origin of these high-energy particles moving with speeds up to 99.9999999999999 per cent of the speed of light, lies in the assumption that they are accelerated by very high electric potentials presumably existing between the giant gas and dust clouds (nebulae) floating in cosmic space. In fact, one could expect that such interstellar clouds would accumulate electric charges in a way similar to the ordinary thunderclouds in our atmosphere, and that the electric potential differences thus created would be much higher than those responsible for the phenomenon of lightning striking between the clouds during thunderstorms.

speeds hit the upper layers of the atmosphere. Passing close by the nuclei of the atoms that form the atmosphere, the primary high-speed electron gradually loses its original energy, which is emitted in the form of gamma radiation all along its track (Figure

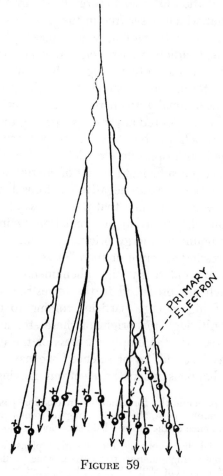

FIGURE 59

The origin of a cosmic ray shower.

59). This radiation gives rise to numerous processes of pair creation, and the newly formed positive and negative electrons rush along the path of the primary particle. Having still a very high energy these secondary electrons give rise to more gamma radiation, which, in its turn, produces still more new electron pairs. This process of successive multiplication is repeated many times during the passage through the atmosphere, so that the primary electron finally arrives at sea level being accompanied by a swarm of secondary electrons half of them positive, the other half negative. It goes without saying that such cosmic-ray showers can also be produced when fast electrons pass through massive material bodies, where, due to the higher density, the branching processes occur with much higher frequency (See Plate IIA).

Turning our attention now to the possible existence of negative protons, we should expect that this kind of particle could be formed by a neutron which had either acquired a negative charge or, what is the same, lost a positive one. It is easy to understand, however, that such negative protons, no more than positive electrons, would be able to exist very long within any ordinary material. In fact, they will be immediately attracted and absorbed by the nearest positively charged atomic nuclei, and most probably will be turned into neutrons after entering the nuclear structure. Thus, if such protons, which would contribute to the symmetry of the present chart of elementary particles, do actually exist in matter, it would not be an easy job to detect them. Remember that positive electrons were found almost half a century after the notion of an ordinary negative electron was introduced into science. Assuming the possible existence of negative protons, we can contemplate the atoms and molecules that are on the—so to say—inverted scheme. Their nuclei, being built from ordinary neutrons and *negative* protons, would have to be surrounded by the envelopes of *positive* electrons. These "inverted" atoms will have properties exactly identical with the properties of ordinary atoms and there will be no way of telling the difference between the inverted water, inverted butter, and so on, and ordinary substances under the same name. There will be no way of telling the difference—unless we bring the ordinary and "inverted" material together. Once, however, two such opposite substances are brought together the processes of mutual annihilation of the oppositely charged electrons, along with mutual neutralization of oppositely charged nucleons will immediately take place, and the mixture will explode with the violence surpassing that of the atomic bomb. For all we know, there may be stellar systems other than ours that are built from such inverted material, in which case any ordinary rock thrown from our system to one constructed the other way, or vice versa, will turn into an atomic bomb as soon as it lands.

At this point we must leave these somewhat fantastic speculations about inverted atoms, and consider still another kind of elementary particles, which, being probably no less unusual, have the merit of actually participating in various observable physical processes—the so-called "neutrinos" which came into physics "through the back door" and, in spite of the "cries of Boeotians" rising against them from many quarters, now occupy an unshakable position in the family of elementary particles. How they were found and recognized constitutes one of modern science's most exciting detective stories.

The existence of *neutrinos* was discovered by a method that a mathematician would call "reductio ad absurdum." The exciting discovery began, not with the fact that something was there, but rather that something was missing. The missing thing was energy, and since energy, according to one of the oldest and most stable laws of physics, can be neither created nor destroyed, discovery that energy that should have been present was absent indicated that there must have been a thief, or a gang of thieves, who took it away. And so the detectives of science, having orderly minds that like to give names to things even when they cannot see them, called the energy thieves "neutrinos."

But that is a bit ahead of the story. To go back into the facts of the great "energy robbery case": As we have seen before, the nucleus of each atom consists of nucleons, about half of them neutral (neutrons), the rest positively charged with electricity. If the balance between the relative number of neutrons and protons in the nucleus is destroyed, by adding one or several extra neutrons or extra protons,[8] an electric adjustment must necessarily take place. If there are too many neutrons some of them will turn into protons by ejecting a negative electron, which leaves the nucleus. If there are too many protons, some of them will turn

---

[8] This can be done by the method of nuclear bombardment described later in this chapter.

into neutrons emitting a positive electron. Two processes of this kind are illustrated in Figure 60. Such electric adjustments of an atomic nucleus are usually known as the beta-decay process, and electrons emitted from the nucleus are known as beta (β)-particles. Since the internal transformation of a nucleus is a well-defined process, it must always be connected with the liberation of a definite amount of energy, which is communicated to the ejected electron. Thus we should expect that the β-electrons

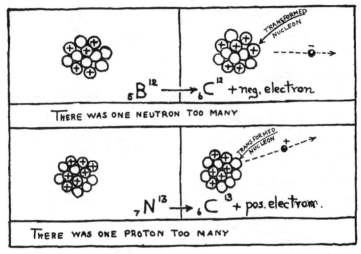

$_5B^{12} \longrightarrow {}_6C^{12} +$ neg. electron

THERE WAS ONE NEUTRON TOO MANY

$_7N^{13} \longrightarrow {}_6C^{13} +$ pos. electron.

THERE WAS ONE PROTON TOO MANY

FIGURE 60

The scheme on negative and positive beta decay (for the convenience of presentation all nucleons are drawn in one plane).

emitted by a given substance must all move with the same velocity. The observational evidence concerning the processes of beta decay stood, however, in a direct contradiction to this expectation. In fact it was found that the electrons emitted by a given substance have different kinetic energies from zero to a certain upper limit. Since no other particles were found and no radiation that would balance this discrepancy, the "case of the missing energy" in the processes of beta decay became quite serious. It was believed for awhile that we were facing here the first experimental evidence of the failure of the famous law of energy conservation, which would be quite a catastrophe for all the elaborate building of physical theory. But there was another possibility: perhaps the missing energy was being carried away by some new kind of particles, which had escaped without having been noticed by any of our observational methods. It was suggested by Pauli that the role of such "Bagdad thieves" of nuclear energy could be played by hypothetical particles called *neutrinos*, which carry no electric charge and whose mass does not exceed the mass of

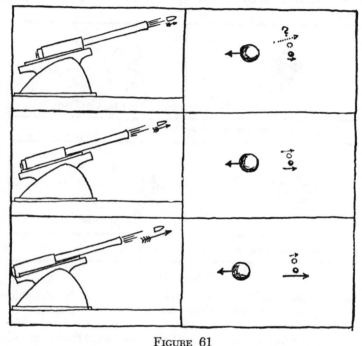

FIGURE 61

The recoil problem in artillery and nuclear physics.

an ordinary electron. In fact, one could conclude from the known facts concerning the interaction of fast-moving particles and matter, that such chargeless, light particles could not be noticed by any existing physical apparatus, and would pass without any difficulty through enormously large thicknesses of any screening material. Thus, whereas the visible light would be completely stopped by a thin metallic filament, and the highly penetrating

X- and gamma radiations would require several inches of lead to be substantially reduced in intensity, a beam of *neutrinos* would go without much difficulty through the thickness of several light-years of lead! No wonder that they escape from any possible observation, and can be noticed only because of the deficit of energy caused by their escape.

But although we cannot catch these neutrinos once they have left the nucleus, there is a way of studying the secondary effect caused by their departure. When you shoot a rifle it hits back against your shoulder, and a big gun rolls back in its carriage after ejecting a heavy shell. The same effect of mechanical recoil is to be expected from atomic nuclei shooting out fast particles, and, in fact, it was observed that the beta-decaying nuclei always acquire a certain velocity in the direction away from the ejected electron. The peculiar property of this nuclear recoil lies however in the observed fact, that no matter whether a fast or a slow electron is being ejected, the recoil velocity of the nucleus is always about the same (Figure 61). This seems very strange since we would naturally expect a fast projectile to produce a stronger recoil in a gun than a slow one. The explanation of the riddle lies in the fact that along with the electron the nucleus always emits a neutrino, which carries the balance of energy. If the electron moves rapidly, taking most of the available energy, the neutrino moves slowly, and vice versa, so that the observed recoil of the nucleus is always strong, owing to the combined effect of *both* particles. If this effect does not prove the existence of the neutrino nothing ever will!

We are ready now to sum up the results of the foregoing discussion and to present a complete list of the elementary particles participating in the structure of the universe, and the relationships that exist among them.

First of all we have the nucleons, which represent the basic material particles. They are, so far as the present state of knowledge can say, either neutral or positively charged, but it is possible that some are negatively charged.

Then we have the electrons representing the free charges of positive and negative electricity.

There are also the mysterious neutrinos, which carry no electric charge and are presumably considerably lighter than electrons.[9]

Finally there are the electromagnetic waves, which account for the propagation of electric and magnetic forces through empty space.

All these fundamental constituents of the physical world are interdependent and can combine in various ways. Thus a neutron can go into a proton by emitting a negative electron and a neutrino (neutron → proton + neg. electron + neutrino); and a proton can go back into a neutron by emitting a positive electron and a neutrino (proton → neutron + pos. electron + neutrino). Two electrons with opposite electrical charges can be transformed into electromagnetic radiation (pos. electron + neg. electron → radiation) or on the contrary can be formed from the radiation (radiation → pos. electron + neg. electron). Finally the neutrinos can combine with electrons, forming the unstable units observed in the cosmic rays and known as *mesons* or, rather incorrectly, as "heavy electrons" (neutrino + pos. electron → pos. meson; neutrino + neg. electron → neg. meson; neutrino + pos. electron + neg. electron → neutral meson).

Combinations of neutrinos and electrons are overloaded with internal energy that makes them about a hundred times heavier than the combined mass of their constituent particles.

Figure 62 shows a schematic chart of elementary particles participating in the structure of the universe.

"But is this the end?" you may ask. "What right have we to assume that nucleons, electrons, and neutrinos are really elementary and cannot be subdivided into still smaller constituent parts? Wasn't it assumed only a half a century ago that the atoms were indivisible? Yet what a complicated picture they present today!" The answer is that, although there is, of course, no way to predict the future development of the science of matter, we have now much sounder reasons for believing that our elementary particles are actually the basic units and cannot be subdivided further. Whereas allegedly indivisible atoms were known to show a great variety of rather complicated chemical, optical, and other proper-

[9] The latest experimental evidence on this subject indicates that a neutrino weighs no more than one tenth as much as an electron.

ties, the properties of elementary particles of modern physics are extremely simple; in fact they can be compared in their simplicity to the properties of geometrical points. Also, instead of a rather large number of "indivisible atoms" of classical physics, we are now left with only three essentially different entities: nucleons, electrons, and neutrinos. And in spite of the greatest desire and

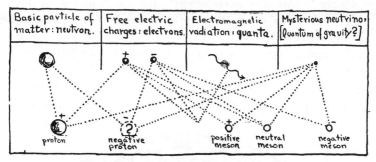

FIGURE 62

The chart of elementary particles of modern physics and of their different intercombinations.

effort to reduce everything to its simplest form, one cannot possibly reduce something to nothing. Thus it seems that we have actually hit the bottom in our search for the basic elements from which matter is formed.

2. THE HEART OF THE ATOM

Now that we have become thoroughly acquainted with the nature and properties of the elementary particles participating in the structure of matter, we may turn to a more detailed study of the nucleus, the heart of every atom. Whereas the structure of the outer body of the atom can be to a certain extent compared to a miniature planetary system, the structure of the nucleus itself presents an entirely different picture. It is clear first of all that the forces holding the nucleus together are not of a purely electric nature, since one half of the nuclear particles, the neutrons, do not carry any electric charge, whereas another half, the protons, are all positively charged, thus repelling each other. And you cannot possibly get a stable group of particles if there is nothing but repulsion between them!

Thus in order to understand why the constituent parts of the nucleus stay together one must necessarily assume that there exist between them forces of some other kind, attractive in nature, which act on uncharged nucleons as well as on the charged ones. Such forces, which, irrespective of the nature of particles involved, make them stay together are generally known as "cohesive forces," and are encountered, for example, in ordinary liquids, where they prevent separate molecules from flying apart in all directions.

In the atomic nucleus we have similar cohesive forces acting between the separate nucleons, and preventing the nucleus from breaking up under the action of electric repulsion between the protons. Thus, in contrast to the outer body of the atom, where

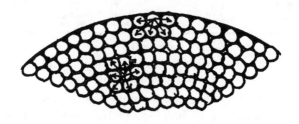

FIGURE 63

Explanation of surface-tension forces in a liquid.

the electrons forming various atomic shells have plenty of space in which to move about, the picture of the nucleus is that of a large number of nucleons packed as tightly together as sardines in a can. As it was first suggested by the author of this book, one may assume that the material of the atomic nucleus is built along the same lines as any ordinary liquid. And just as in the case of ordinary liquids, we have here the important phenomenon of surface tension. It may be remembered that the phenomenon of surface tension in liquids arises from the fact that, whereas a particle located in the interior is pulled equally in all directions by its neighbors, the particles located on the surface are subject to the forces that attempt to pull them inwards (Figure 63).

This results in the tendency of any liquid droplet not subject to

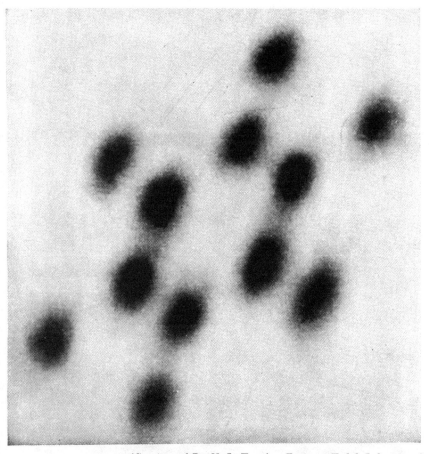

PLATE I

Photograph of hexamethylbenzene molecule magnified 175,000,000 times.

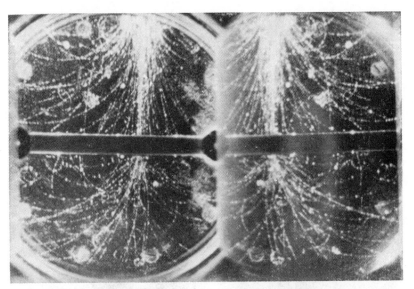

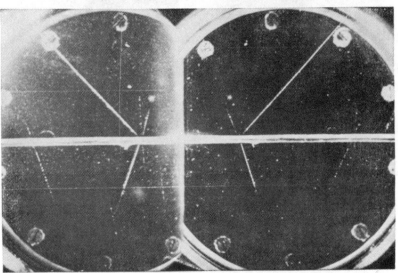

(Photographed by Carl Anderson, California Institute of Technology.)
PLATE II

A. Cosmic-ray shower originating in the outer wall of the cloud chamber, and again in the lead plate in the middle. Positive and negative electrons forming the shower are deflected in opposite directions by magnetic field.

B. Nuclear disintegration produced by cosmic ray particle in the middle plate.

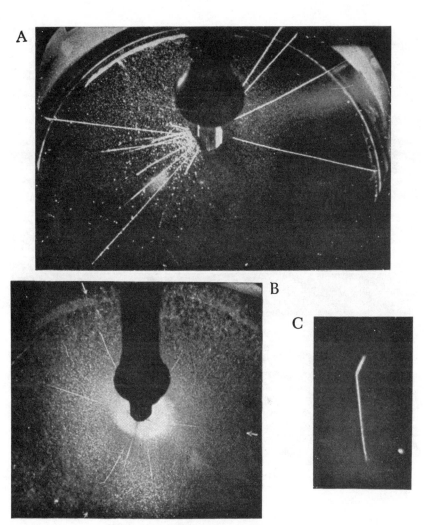

(*Photographed by Drs. Dee and Feather in Cambridge.*)

PLATE III

Transformations of atomic nuclei caused by artificially accelerated projectiles.

A. A fast deuteron hits another deuteron from the heavy hydrogen gas in the chamber, producing the nuclei of tritium and ordinary hydrogen ($_1D^2 + _1D^2 \rightarrow _1T^3 + _1H^1$).

B. A fast proton hits the nucleus of boron, breaking it into three equal parts ($_5B^{11} + _1H^1 = 3 \, _2He^4$).

C. A neutron coming from the left and invisible in the picture breaks the nucleus of nitrogen into a nucleus of boron (upward track) and a nucleus of helium (downward track). ($_7N^{14} + _0n^1 \rightarrow _5B^{11} + _2He^4$).

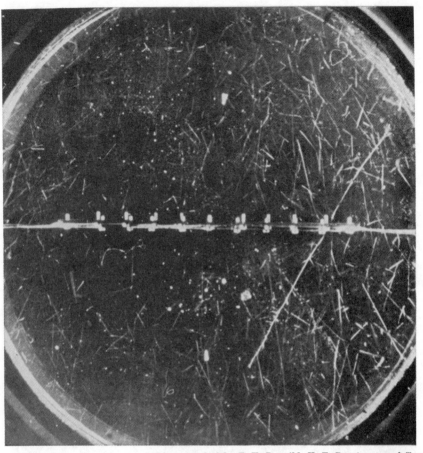

(*Photographed by T. K. Boggild, K. T. Brostrom, and Tom
Lauritsen at the Institute of Theoretical Physics in Copenhagen.*)

PLATE IV

A cloud-chamber photograph of the fission of a uranium nucleus. A
neutron (which is, of course, not seen in the picture) hits one of the
uranium nuclei in a thin layer placed across the chamber. The two
tracks correspond to two fission fragments flying apart with the energy
of about 100 Mev each.

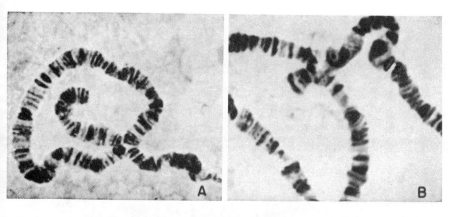

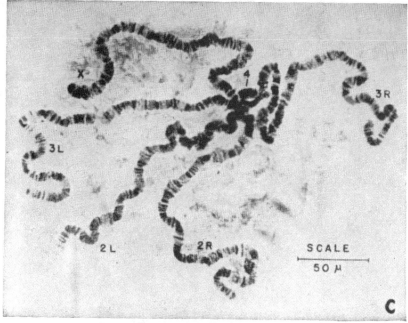

(From Drosophila Guide, *by M. Demerec and B. P. Kaufmann.
Washington, Carnegie Foundation of Washington, 1945. Used
by permission of Mr. Demerec.)*

PLATE V

A and B. Photomicrographs of salivary-gland chromosomes of *D.
melanogaster*, showing inversion and reciprocal translocation.
C. Photomicrograph of female larva of *D. melanogaster*. X, the X
chromosomes, closely paired, side by side; 2L and 2R, the left and
right limb of the paired second chromosomes; 3L and 3R, the third
chromosomes; 4, the fourth chromosomes.

PLATE VI

Living molecules? The particles of tobacco mosaic virus magnified 34,800 times. This picture was taken by means of an electron microscope.

PLATE VII

A. Spiral nebula in Ursa Major, a distant island universe, seen from above.

B. The spiral nebula in Coma Berenices, another distant island universe, seen on edge.

(Mt. Wilson Observatory photographs.)

PLATE VIII

The Crab Nebula. An expanding envelope of gases thrown out by a supernova observed at this place of the sky in the year 1054 by Chinese astronomers.

If the rope holds!

"Deimos"

Figure 64

any outside forces to assume a spherical shape, since a sphere is the geometrical figure possessing the smallest surface for any given volume. Thus we are led to the conclusion that *atomic nuclei of different elements may be considered simply as variously sized droplets of a universal "nuclear fluid."* We must not forget, however, that nuclear fluid, though qualitatively very similar to ordinary liquids, is rather different quantitatively. In fact its density exceeds the density of water by a factor of 240,000,000,000,000, and its surface tension forces are about 1,000,000,000,000,000,000 times larger than those of water. To make these tremendously large numbers more understandable, let us consider the following example. Suppose we have a wire frame roughly in the shape of an inverted capital U, about 2 in. square as shown full size in Figure 64, with a piece of straight wire across it, and with a soap film across the square thus formed. The surface tension forces of the film will pull the crossbar wire upwards. We may counteract these surface-tension forces by hanging a little weight on the crossbar. If the film is made of ordinary water with some soap dissolved in it, and is say 0.01 mm thick, it will weigh about 1/4 g, and will support a total weight of about 3/4 g.

Now, if it were possible to make a similar film from nuclear fluid, the total weight of the film would be fifty million tons (about the weight of one thousand ocean liners), and we could hang on the cross wire a load of about a thousand billion tons, which is roughly the mass of "Deimos," the second satellite of Mars! One would have to have rather powerful lungs in order to be able to blow a soap bubble from nuclear fluid!

Considering atomic nuclei as tiny droplets of nuclear fluid, we must not overlook the important fact that these droplets are electrically charged, since about one half of the particles forming the nucleus are protons. The forces of electric repulsion between nuclear constituent particles trying to disrupt the nucleus into two or more parts are counteracted by the surface tension forces that tend to keep it in one piece. Here lies the principal reason for the instability of atomic nuclei. If the surface-tension forces prevail, the nucleus will never break up by itself, and two nuclei, coming into contact with each other, will have a tendency to *fuse* just as two ordinary droplets do.

If, on the contrary, the electric forces of repulsion have the upper hand, the nucleus will show a tendency to break spontaneously into two or more parts, which will fly apart at high speed; such a breaking-up process is usually designated by the term "fission."

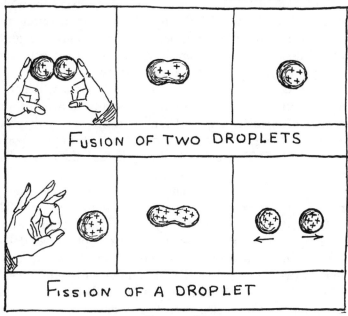

FIGURE 65

Exact calculations concerning the balance between the surface-tension and electric forces in the nuclei of different elements, were made by Bohr and Wheeler (in 1939) and led to the extremely important conclusion that, whereas the surface-tension forces hold the upper hand in the nuclei of all the elements in the first half of the periodic system (approximately up to silver), the electric repulsive forces prevail for all heavier nuclei. Thus the nuclei of all elements heavier than silver are principally unstable, and, under the action of a sufficiently strong fillip from outside, would break up into two or more parts, with the liberation of a considerable amount of internal nuclear energy (Figure 65*a*). On the contrary, we should expect a spontaneous fusion process whenever two light nuclei with a combined atomic weight less than that of silver come close together (Figure 65*b*).

It must be remembered, however, that neither the fusion of two light nuclei, nor the fission of a heavy nucleus would normally take place unless we did something about it. In fact, to cause the fusion of two light nuclei we have to bring them close together against the repulsive forces interacting between their charges, and in order to force a heavy nucleus to go through the process of fission we must start it vibrating with a sufficiently large amplitude, by giving to it a strong tap.

This state of affairs, in which a certain process will not get under way without initial excitation, is generally known in science as *the state of metastability*, and can be illustrated by the exam‧ples of a rock hanging over a precipice, a match in your pocket, or a charge of TNT in a bomb. In each case there is a large amount of energy waiting to be set free, but the rock will not roll down unless kicked, the match will not burn unless heated by friction against your shoe sole or something else, and TNT will not explode unless detonated by a fuse. The fact that we live in a world in which practically every object except a silver dollar[10] is a potential nuclear explosive, without being blown to bits, is due to the extreme difficulties that attend the starting of a nuclear reaction, or in more scientific language, to the extremely high activation energies of nuclear transformations.

In respect to nuclear energy we live (or rather lived until quite recently) in a world similar to that of an Eskimo dwelling in a subfreezing temperature for whom the only solid is ice and the only liquid alcohol. Such an Eskimo would never have heard about fire, since one cannot get fire by rubbing two pieces of ice against each other, and would consider alcohol as nothing but a pleasant drink, since he would have no way of raising its temperature to the burning point.

And the great perplexity of humanity caused by the recently discovered process of liberating on a large scale the energy hidden in the interior of the atom can be compared to the aston-

[10] It will be remembered that silver nuclei will neither fuse nor fission.

ishment of our imaginary Eskimo when shown an ordinary alcohol burner for the first time.

Once the difficulty of starting a nuclear reaction is overcome, however, the results would pay proportionately for all the troubles involved. Take, for example, a mixture of equal amounts of oxygen and carbon atoms. Uniting chemically according to the equation

$$O + C \rightarrow CO + energy,$$

these substances would give us 920 calories[11] per gram of mixture.

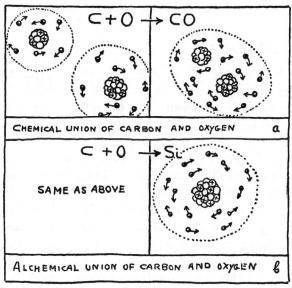

FIGURE 66

If, instead of ordinary chemical union (molecular fusion) (Figure 66a) between these two atomic species we have an alchemical union (nuclear fusion) between their nuclei (Figure 66b):

$$_6C^{12} + {_8}O^{16} = {_{14}}Si^{28} + energy$$

the energy liberated per gram of mixture will be 14,000,000,000 calories, that is, 15,000,000 times as great.

Similarly the breaking up of a complex TNT molecule into the

[11] A calory is a unit of heat defined as the energy necessary to raise 1 gram of water 1 degree centigrade.

molecules of water, carbon monoxide, carbon dioxide, and nitrogen (molecular fusion) liberates about 1000 calories per gram, whereas an equal weight of, let us say, mercury undergoing the process of nuclear fission would give us altogether 10,000,000,000 cal.

It must not be forgotten however that whereas most chemical reactions would take place easily at temperatures of a few hundred degrees, corresponding nuclear transformations would not even start before the temperature reaches many millions of degrees! This difficulty of starting nuclear reaction accounts for the comforting fact that there is no immediate danger that the entire universe will turn into pure silver in one tremendous explosion.

3. *ATOM SMASHING*

Although the integrity of atomic weights represents a very strong argument in favor of the complexity of atomic nuclei, the final proof of such complexity can be achieved only by direct empirical evidence concerning the possibility of breaking a nucleus into two or more separate parts.

The first indication that such a break-up process can really take place was supplied fifty years ago (in 1896) by Becquerel's discovery of radioactivity. It was, in fact, shown that the highly penetrating radiation (similar to ordinary X rays), that is emitted spontaneously by the atoms of such elements as uranium and thorium located near the upper end of the periodic system is due to the slow spontaneous decay of these atoms. The careful experimental study of this newly discovered phenomenon led soon to the conclusion that the decay of a heavy nucleus consists in its spontaneous breaking up into two largely unequal parts: (1) a small fragment, known as an *alpha particle,* representing the atomic nucleus of helium, and (2) the remainder of the original nucleus, which represents the nucleus of the daughter element. When the original uranium nucleus breaks up, ejecting α-particles, the resulting nucleus of the daughter element known as *Uranium X_I* undergoes an internal electric readjustment emitting two free charges of negative electricity (ordinary electrons) and turning into the nucleus of uranium isotope, which is four units lighter than the original uranium nucleus. This electric adjustment is followed again by a series of emissions of α-particles, then by more electrical adjustments, etc., until we come finally to the nucleus of the lead atom, which appears stable, and does not decay.

A similar series of successive radioactive transformations with the alternate emission of α-particles, and electrons is observed in two other radioactive families: The thorium-family starting with the heavy element thorium, and the actinium-family starting with the elements known as actino-uranium. In these three families the processes of spontaneous decay continue until there are left only the three different isotopes of lead.

An inquisitive reader will probably be surprised in comparing the above description of spontaneous radioactive decay with the general discussion of the previous section, in which it was stated that *the instability of an atomic nucleus must be expected in all the elements of the second half of the periodic system,* where the disruptive electric forces have the upper hand over the forces of surface tension tending to hold the nucleus in one piece. If all the nuclei heavier than silver are unstable, why then is the spontaneous decay observed only for a few of the heaviest elements such as uranium, radium, and thorium? The answer is that, theoretically speaking, *all* elements heavier than silver must be considered as radioactive elements, and are as a matter of fact slowly being transformed by decay into the lighter elements. But in most cases the spontaneous decay takes place so very slowly that there is no way of noticing it. Thus in such familiar elements as iodine, gold, mercury, and lead the atoms may break up at the rate of one or two in many centuries, which is too slow to be recorded even by the most sensitive physical instruments. Only in the heaviest elements is the tendency to break up spontaneously strong enough to result in noticeable radioactivity.[12] The com‧ parative transformation rates also govern the way in which a given unstable nucleus breaks up. Thus the nucleus of a uranium atom, for example, can break up in many different ways: it may split spontaneously into two equal parts, or into three equal parts, or into several parts of widely varying sizes. However, the

[12] In uranium, for example, we have several thousand atom breakings per second in each gram of material.

easiest way for it to divide is into an α-particle and the remaining heavy part, and that is why it usually happens this way. It has been observed that the spontaneous break-up of a uranium nucleus into two halves is about a million times less probable than the chipping off of an α-particle. Thus, whereas in one gram of uranium some ten thousand nuclei break up each second by each emitting an α-particle, we shall have to wait for several minutes to see a spontaneous fission process in which a uranium nucleus breaks into two equal halves!

The discovery of the radioactive phenomena proved beyond any doubt the complexity of nuclear structure and paved the way for the experiments on the artificially produced (or induced)

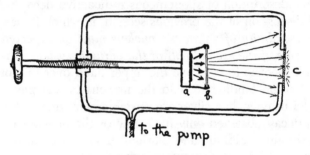

FIGURE 67

How the atom was split the first time.

nuclear transformations. This question then arose: if the nuclei of heavy, particularly unstable, elements decay on their own initiative, can we not break up the nuclei of other ordinarily stable elements by hitting them hard enough with some rapidly moving nuclear projectile?

With this thought in mind, Rutherford decided to subject the atoms of various ordinarily stable elements to an intense bombardment by the nuclear fragments (the α-particles) resulting from the spontaneous breaking up of unstable radioactive nuclei. The apparatus used in 1919 by Rutherford in his first experiments in nuclear transformations (Figure 67) is the acme of simplicity compared with the giant atom smashers used nowadays in several physics laboratories. It consisted of an evacuated cylindrical vessel with a thin window made from a fluorescent material (*c*), which acted as a screen. The source of bombarding α-particles was a thin layer of radioactive substance deposited on the metallic plate (*a*), and the element to be bombarded (aluminum in this case) was in the form of a thin filament (*b*) placed some distance away from the source. The target filament was arranged in such a way that all incident α-particles would remain embedded in it, once they had encountered it, so that it would be impossible for them to illuminate the screen. Thus the screen would remain completely dark unless it was affected by secondary nuclear fragments emitted from the target material as a result of the bombardment.

Putting everything in its place and looking at the screen through a microscope Rutherford saw a sight that could hardly be mistaken for darkness. The screen was alive with myriads of tiny sparks flashing here and there over its entire surface! Each spark was produced by the impact of a proton against the material of the screen, and each proton was a "fragment" kicked out from an aluminum atom in the target by the incident α-projectile. Thus the theoretical possibility of an artificial transformation of elements became a scientifically established fact.[13]

During the decades immediately following Rutherford's classic experiment, the science of artificial transformation of elements became one of the largest and one of the most important branches of physics, and tremendous progress was achieved in methods both of producing fast projectiles for the purpose of nuclear bombardment, and of observing the obtained results.

The instrument that most satisfactorily permits us to see with our own eyes what happens when a nuclear projectile hits a nucleus is known as a cloud chamber (or Wilson chamber after its inventor). It is represented schematically in Figure 68. Its operation is based on the fact that fast-moving charged particles, such as α-particles, produce on their way through the air, or through any other gas, a certain distortion in the atoms situated along their route. With their strong electric fields, these projectiles tear off one electron or more from the atoms of gas that happen to be in their way, leaving behind a large number of

---

[13] The process described above may be represented by the formula:

$$_{13}Al^{27} + _2He^4 \rightarrow _{14}Si^{30} + _1H^1.$$

ionized atoms. This state of affairs does not last very long, for very soon after the passage of the projectile the ionized atoms will catch back their electrons, returning to the normal state. But if the gas in which such ionization takes place is saturated with water vapor, tiny droplets will be formed on each of the ions—it is a property of water vapor that it tends to accumulate on ions, dust particles, and so on—producing a thin band of fog along the track of the projectile. In other words, the track of any

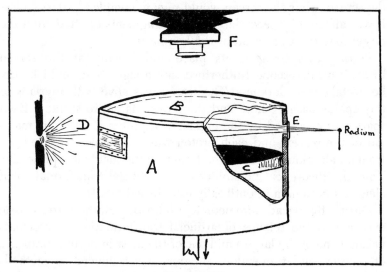

FIGURE 68
The scheme of Wilson's cloud-chamber.

charged particle moving through a gas thus becomes visible in the same way as does the track of a smoke-writing airplane.

From the technical point of view, the cloud chamber is a very simple apparatus, consisting essentially of a metallic cylinder (*A*) with a glass cover (*B*) containing a piston (*C*), which can be moved up and down by an arrangement not shown in the picture. The space between the glass cover and the surface of the piston is filled with ordinary atmospheric air (or any other gas, if so desired) containing a considerable amount of water vapor. If the piston is abruptly pulled down immediately after some atomic projectiles have entered the chamber through the window (*E*) the air above the piston will cool and the water vapor will begin to precipitate, in the form of thin bands of fog, along the track of the projectiles. These bands of fog, being illumined by a strong

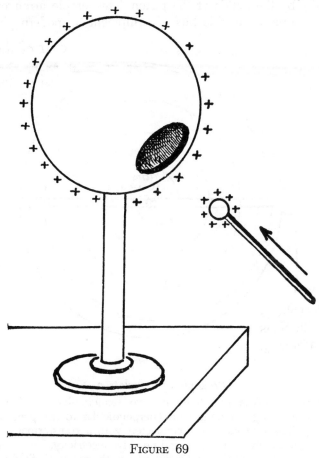

FIGURE 69

Principle of the electrostatic generator

It is well known from elementary physics that a charge communicated to a spherical metallic conductor is distributed on its surface. Thus we can charge such a conductor to arbitrarily high potentials by introducing, one by one, small charges into its interior by bringing a small charged conductor through a hole made in the sphere and touching its surface *from inside*. In practice one uses actually a continuous belt entering into the spherical conductor through the hole and carrying in electric charges produced by a small transformer.

light through a side window (*D*), will stand out clearly against the blackened surface of the piston and can be observed visually or photographed by the camera (*F*), which is operated automatically by the action of the piston. This simple arrangement, one of the most valuable bits of equipment in modern physics,

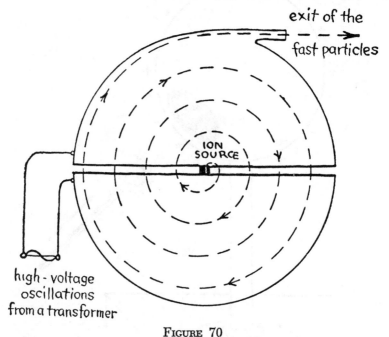

FIGURE 70

Principle of a cyclotron

A cyclotron consists essentially of two semicircular metallic boxes placed in a strong magnetic field (perpendicular to the plane of the drawing). The two boxes are connected with a transformer and are charged alternately by positive and negative electricity. The ions coming from the source in the center describe in the magnetic field circular trajectories accelerated each time they pass from one box into the other. Moving faster and faster, the ions describe an unwinding spiral, and finally come out at a very high speed.

permits us to obtain beautiful photographs of the results of nuclear bombardment.

It was also naturally desirable to devise methods by which one could produce strong beams of atomic projectiles simply by accelerating various charged particles (ions) in strong electric fields. Apart from removing the necessity of using rare and expensive radioactive substances, such methods permit us to use other different types of atomic projectiles (as for example protons), and to attain kinetic energies higher than those supplied by ordinary radioactive decay. Among the most important machines for producing intensive beams of fast moving atomic projectiles are the *electrostatic generator,* the *cyclotron,* and the *linear accelerator,* represented with short descriptions of their functioning in Figures 69, 70 and 71 respectively.

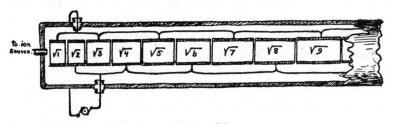

FIGURE 71
Principle of a linear accelerator

This arrangement consists of a number of cylinders of increasing length that are being charged alternately positively and negatively by a transformer. Passing from one cylinder into another the ions are gradually accelerated by the existing potential difference, so that their energy increases each time by a given amount. Since the velocity is proportional to the square root of the energy, the ions will be kept in phase with the alternating field, if the length of cylinders is proportional to the square roots in integer numbers. Building a sufficiently long system of this type we can accelerate the ions to any desired speed.

Using the above described types of electric accelerators for producing powerful beams of various atomic projectiles, and directing these beams against targets made from different materials, we can obtain a large number of nuclear transformations, which can be conveniently studied by means of cloud-chamber photographs. Some of these photographs, showing the individual processes of nuclear transformations, are shown in Plates III and IV.

The first picture of this kind was taken by P. M. S. Blackett in Cambridge, and represented a beam of natural α-particles passing through a chamber filled with nitrogen.[14] It showed first of all that the tracks have a definite length, owing to the fact that, flying through the gas, the particles gradually lose their kinetic energy, coming ultimately to a stop. There were two distinctly different groups of track lengths corresponding to the two groups of α-particles with different energies present in the source (a mixture of two alpha-emitting elements: ThC and ThC¹). One could notice that, being in general quite straight, α-tracks show well-defined deflections near the end where the particles have lost most of their initial energy and can be more easily deflected by indirect collision with the nuclei of nitrogen atoms that they encounter on their way. But the star feature of this photograph lay in one particular α-track, which showed a characteristic branching, one branch being long and thin, another short and thick. It showed the result of a direct head-on collision between the incident α-particle and the nucleus of one of the nitrogen atoms in the chamber. The thin long track represented the trajectory of the proton knocked out of the nitrogen nucleus by the force of the impact, whereas the short thick track corresponded to the nucleus itself thrown aside in the collision. The fact that there was no third track that would correspond to the ricocheted α-particle, indicated that the incident α-particle had adhered to the nucleus and was moving together with it.

In Plate IIIв we see the effect of artificially accelerated protons colliding with the nuclei of boron. The beam of fast protons issuing from the accelerator's nozzle (dark shadow in the middle of the photograph) hits a layer of boron placed against the opening, and sends nuclear fragments flying in all directions through the surrounding air. An interesting feature of this photograph is that the fragment's tracks appear always in triplets (two such triplets, one marked with arrows, can be seen in the photograph), because the nucleus of boron, being hit by a proton, breaks up into three equal parts.[15]

Another photograph, Plate IIIА, shows collisions between the fast-moving deuterons (the nuclei of heavy hydrogen formed

[14] The alchemic reaction recorded on Blackett's photograph (not reproduced in this book) is represented by the equation: $_7N^{14} + _2He^4 \rightarrow _8O^{17} + _1H^1$.

[15] The equation of this reaction is: $_5B^{11} + _1H^1 \rightarrow _2He^4 + _2He^4 + _2He^4$.

by one proton and one neutron) and other deuterons in the target material.[16]

The longer tracks seen in the picture correspond to protons ($_1H^1$-nuclei) whereas the shorter ones are due to the nuclei of triple-heavy hydrogen known as tritons.

No cloud-chamber picture gallery would be complete without the nuclear reaction involving the neutrons, which, together with protons, constitute the main structural elements of every nucleus.

It would be quite futile to look for neutron tracks in the cloud-chamber pictures, since, having no electric charge, these "dark horses of nuclear physics" pass through matter without producing any ionization whatsoever. But when you see the smoke from a hunter's gun, and the duck falling down from the sky, you know there *was* a bullet even though you cannot see it. Similarly looking at the cloud-chamber photograph, Plate IIIc, which shows a nucleus of nitrogen breaking up into helium (downward track) and boron (upward track), you cannot help feeling that this nucleus was hard hit by some invisible projectile coming from the left. And, indeed, in order to get such a photograph one has to place at the left wall of the cloud chamber a mixture of radium and beryllium, which is known to be a source of fast neutrons.[17]

The straight line along which the neutron was moving through the chamber can be seen at once by connecting the position of the neutron source with the point where the breaking up of the nitrogen atom took place.

The fission process of the uranium nucleus is shown in Plate IV. This photograph was taken by Boggild, Brostrom, and Lauritsen and shows two fission fragments flying in opposite directions from a thin aluminum foil supporting the bombarded uranium layer. Neither the neutron which produced the fission, nor the neutrons resulting from it would, of course, show on the picture. We could go on indefinitely describing various types of nuclear transformations obtainable by the method of nuclear bombard-

[16] This reaction is represented by the equation: $_1H^2 + _1H^2 \rightarrow _1H^3 + _1H^1$.

[17] In terms of alchemic equation the processes that take place here can be written in the following form: (a) production of the neutron: $_4Be^9 + _2He^4$ (α-particle from Ra) $\rightarrow _6C^{12} + _0n^1$; (b) neutron impact against nitrogen nucleus: $_7N^{14} + _0n^1 \rightarrow _5B^{11} + _2He^4$.

ment by electrically accelerated projectiles, but it is the time now to turn to a more important question concerning *the efficiency* of such bombardment. It must be remembered that the pictures shown in Plates III and IV represent individual cases of the disintegration of single atoms, and that in order to turn, let us say, one gram of boron completely into helium we should have to break every one of the 55,000,000,000,000,000,000,000 atoms contained in it. Now, the most powerful electric accelerator produces about 1,000,000,000,000,000 projectiles per second, so that even if every projectile were to break one nucleus of boron we should have to run the machine for 55 million seconds or about two years to finish the job.

The truth is, however, that the effectiveness of charged nuclear projectiles produced in various accelerating machines is much

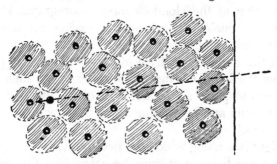

FIGURE 72

smaller than that, and usually only one projectile out of several thousand can be counted upon to produce a nuclear crack-up in the bombarded material. The explanation of this extremely low efficiency of atomic bombardment lies in the fact that atomic nuclei are surrounded by the envelopes of electrons that have the power to slow down the charged atomic projectiles moving through them. Since the target area of the atomic envelope is much larger than the target area of the nucleus and since we cannot, of course, aim atomic projectiles directly at the nucleus, each such projectile must necessarily pierce many atomic envelopes before it will have the chance to deliver a direct blow to one of the nuclei. The situation is explained graphically in Figure 72, where atomic nuclei are represented by solid black spheres and their electronic envelopes by lighter shadows. The ratio of atomic and nuclear diameters is about 10,000 so that the target areas stand in the ratio of 100,000,000 to 1. On the other hand, we know that a charged particle passing through an electronic envelope of an atom loses about one hundredth of one per cent of its energy, so that it will be stopped completely after passing through some 10,000 atomic bodies. It is easy to see from the above quoted numbers that only about 1 particle in 10,000 will have a chance to hit the nucleus before all its initial energy has been dissipated in the atomic envelopes. Taking into account this low efficiency of charged projectiles in delivering a destructive blow to the nuclei of the target material we find that in order to transform completely 1 g of boron, we must keep it in the beam of a modern atom-smashing machine for the period of at least 20,000 years!

4. NUCLEONICS

"Nucleonics" is a very inappropriate word, but like many such words it seems to remain a part of practical usage, and there is nothing to be done about it. As the term "electronics" is used to describe knowledge in the broad field of practical application of free electron beams, the term "nucleonics" should be understood to apply to the science of practical applications of nuclear energy liberated on a large scale. We have seen in the previous sections that the nuclei of various chemical elements (except silver) are overloaded with tremendous amounts of internal energy that can be liberated by the processes of nuclear fusion in the case of lighter elements, and by nuclear fission in the case of heavier ones. We have also seen that the method of nuclear bombardment by artificially accelerated charged particles, though of great importance for the theoretical study of various nuclear transformations, cannot be counted upon for practical use because of its extremely low efficiency.

Since the ineffectiveness of ordinary nuclear projectiles, such as α-particles, protons, and so on, lies essentially in their electric charge, which causes them to lose their energy while passing through atomic bodies, and prevents them from coming sufficiently close to the charged nuclei of bombarded material, we must expect that much better results would be obtained by using the uncharged projectiles and bombarding various atomic nuclei with neutrons. Here, however, is the catch! Owing to the fact that neutrons can without any difficulty penetrate the nuclear structure, they do not exist in nature in the free form, and whenever a free neutron is artificially kicked out of some nucleus by an incident projectile (for example a neutron from beryllium nuclei subjected to alpha bombardment) it will very soon be recaptured by some other nucleus.

Thus in order to produce strong beams of neutrons for the purposes of nuclear bombardment, we have to kick out every single one of them from the nuclei of some element. This brings us back to the low efficiency of charged projectiles that must be used for this purpose.

There is, however, one way out of this vicious circle. If it were possible to have neutrons kick out neutrons and to do it in such a way that each neutron would produce more than one offspring, these particles would multiply like rabbits (compare Figure 97) or bacteria in infected tissue, and the descendants of one single neutron would soon become sufficiently numerous to attack every single atomic nucleus in a large lump of material.

The big boom in nuclear physics, which brought it from the quiet ivory tower of pure science concerned with the most intimate properties of matter, into the noisy whirlpool of shouting newspaper headlines, heated political discussions, and stupendous industrial and military developments, is due to the discovery of one particular nuclear reaction that makes such a neutron multiplication process possible. Everybody who reads newspapers knows that nuclear energy, or atomic energy as it is commonly called, can be released through the fission process of uranium nuclei discovered by Hahn and Strassman late in 1938. But it would be a mistake to believe that the fission itself, that is, the splitting of a heavy nucleus into two nearly equal parts, could contribute to the progressive nuclear reaction. In fact, the two nuclear fragments resulting in fission carry heavy electric charges (about a half charge of the uranium nucleus each), which prevent them from approaching too close to other nuclei. Thus, rapidly losing their initially high energy to the electronic envelopes of neighboring atoms, these fragments will rapidly come to rest without producing any further fissions.

What makes the fission process so important for the development of a self-sustaining nuclear reaction is the discovery that before being finally slowed down each fission fragment emits a neutron (Figure 73).

This peculiar aftereffect of fission is due to the fact that, like the two pieces of a broken spring, the two broken halves of a

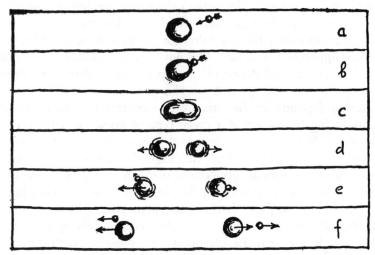

FIGURE 73
Successive stages of the fission process.

heavy nucleus begin their existence in a state of rather violent vibration. These vibrations, which are not able to cause a secondary nuclear fission (of each of the fragments into two), are, however, strong enough to cause the ejection of some nuclear structural units. When we say that each fragment emits one neutron, we mean it only in a statistical sense; in some cases two or even three neutrons may be ejected from a single fragment— while in other cases none. The average number of neutrons emitted from a fission fragment depends, of course, on the intensity of its vibrations, which, in turn, is determined by the total energy release in the original fission process. Since, as we have seen above, the energy set free in fission increases with the weight of the nucleus in question, we must expect that the mean number of neutrons per fission fragment also increases along the periodic system. Thus, the fission of a gold nucleus (which has not yet been achieved experimentally because of the very high initiation energy required in this case) would probably give considerably less than one neutron per fragment; the fission of uranium nuclei gives on the average about one neutron per fragment (about two neutrons per fission); whereas in the fission of still heavier elements (as for example plutonium) the mean number of neutrons per fragment may be expected to be larger than one.

In order to satisfy the condition for progressive neutron breeding it is apparently necessary that out of, say, a hundred neutrons entering into the substance we should get more than a hundred neutrons of the next generation. The possibility of fulfilling this condition depends on the comparative effectiveness of neutrons in producing the fission of a given type of nuclei, and the mean number of fresh neutrons produced in an accomplished fission. It must be remembered that, although the neutrons are much more effective nuclear projectiles than the charged particles, their effectiveness in producing the fission is, however, not a hundred per cent. In fact, there is always a possibility that upon entering the nucleus a high velocity neutron will give to the nucleus only a part of its kinetic energy, escaping with the rest of it; in such cases the energy will be dissipated between several nuclei, none of them getting enough to cause the fission.

It can be concluded from the general theory of nuclear structure that the fission effectiveness of neutrons increases with the increasing atomic weight of the element in question, coming fairly close to a hundred per cent for the elements near the end of the periodic system.

We can now work out two numerical examples corresponding to the favorable and unfavorable conditions for neutron breeding. (a) Suppose we have an element in which the fission efficiency of fast neutrons is 35 per cent and the mean number of neutrons produced per fission is 1.6.[18] In such a case 100 original neutrons

[18] These numerical values are chosen entirely for the sake of an example, and do not correspond to any actual nuclear species.

will produce altogether 35 fissions, giving rise to $35 \times 1.6 = 56$ neutrons of the next generation. It is clear that in this case the number of neutrons will rapidly drop with time, each generation being only about one half of the previous one. (b) Suppose now we take a heavier element in which fission efficiency of neutrons rises to 65 per cent, and the mean number of neutrons produced per fission to 2.2. In this case our 100 original neutrons will pro-

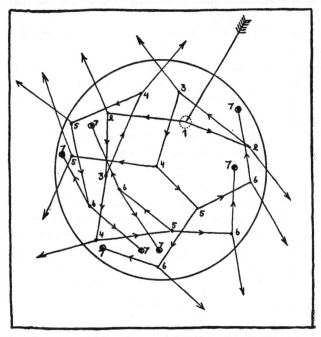

FIGURE 74

A nuclear chain reaction started in a spherical piece of fissionable material by a stray neutron. Although many neutrons are lost by crossing the surface, the number of neutrons in consecutive generations is increasing, leading to an explosion.

duce 65 fissions giving a total of $65 \times 2.2 = 143$. With each new generation the number of neutrons will grow by about 50 per cent, and within a very short time there will be enough of them to attack and break up any single nucleus in the sample. We are considering here the progressive *branching chain reaction,* and call the substances in which such a reaction can take place *fission-able substances.*

A careful experimental and theoretical study of the conditions necessary for the development of progressive branching chain reactions leads to the conclusion that among all the variety of nuclear species existing in nature, *there is only one particular brand of nuclei to which such reaction is possible. These are the nuclei of the famous light isotope of uranium, U-235, the only natural fissionable substance.*

However, U-235 does not exist in nature in a pure form, and is always found to be strongly diluted by the heavier unfission-able isotope of uranium, U-238 (0.7 per cent of U-235 and 99.3 per cent of U-238), which hinders the development of the prog-ressive chain reaction in natural uranium in the very same way as the presence of water prevents wet wood from burning. It is, in fact, only because of this dilution by the inactive isotope that the highly fissionable atoms of U-235 still exist in nature, since otherwise they would have been all destroyed long ago by a fast chain reaction among them. Thus, in order to be able to use the energy of U-235 one must either separate these nuclei from the heavier nuclei of U-238, or one must devise the method for neutralizing the disturbing action of the heavier nuclei without actually removing them. Both methods were actually followed in the work on the problem of atomic energy liberation, leading in both cases to successful results. We shall discuss them here only briefly, since technical problems of such kind do not fall within the scope of the present book.[19]

The straightforward separation of the two uranium isotopes represents a very difficult technical problem, since, owing to their identical chemical properties, such a separation cannot be achieved by the ordinary methods of industrial chemistry. The only difference between these two kinds of atoms lies in their masses, one being 1.3 per cent heavier than the other. This sug-gests the separation methods based on such processes as diffusion, centrifuging, or the deflection of ion beams in magnetic and

[19] For more detailed discussion the reader is referred to the book by Selig Hecht, *Explaining the Atom,* first published by Viking Press in 1947. A new edition, revised and expanded by Dr. Eugene Rabinowitch, is avail-able in the Explorer paperbound series.

electric fields, where the mass of the separate atoms plays the predominant rôle. In Figure 75 *a, b,* we give a schematic presentation of the two major separation methods with a short description of each.

The disadvantage of all these methods lies in the fact that, owing to a small difference of mass between the two uranium isotopes, the separation cannot be achieved in one single step, but requires a large number of repetitions, which lead to the

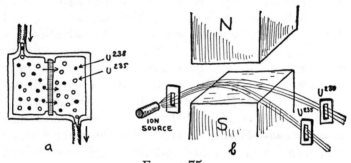

FIGURE 75

(*a*) Separation of isotopes by the diffusion method. The gas containing both isotopes is pumped into the left part of the chamber and diffuses through the wall separating it from the other part. Since light molecules diffuse faster the fraction on the right becomes enriched in U-235.

(*b*) Separation of isotopes by the magnetic method. The beam is sent through a strong magnetic field, and the molecules containing the lighter U-isotope are deflected more strongly. Since to have a good intensity one must use wide slits, the two beams (with U-235 and U-238) partially overlap and we get again only partial separation.

products more and more enriched in the light isotope. However, after a sufficient number of repetitions, reasonably pure samples of U-235 can finally be obtained.

A much more ingenious method consists in running the chain reaction in natural uranium in which the disturbing action of the heavier isotope is artificially reduced by the use of the so-called moderator. In order to understand this method we must remember that the negative effect of the heavier uranium isotope consists essentially in absorbing a large percentage of neutrons produced in U-235 fissions, thus cutting off the possibility for the development of a progressive chain reaction. Thus, if we could do something to prevent the nuclei of U-238 from kidnapping the neutrons before they have a chance to meet a U-235 nuclei, which would cause their fission, the problem would be solved. At first sight the task of preventing the U-238 nuclei, which are 140 times more numerous than U-235 nuclei, from getting the lion's share of neutrons seems to be quite impossible. We are, however, helped in this problem by the fact that the "neutron-capture-ability" of the two uranium isotopes is different depending on the speed with which the neutron is moving. For fast neutrons, as they come from the fissioning nucleus, the capture-abilities of both isotopes are the same, so that U-238 will capture 140 neutrons for each neutron captured by U-235. For the neutrons of intermediate speeds U-238 nuclei are somewhat better catchers than the nuclei of U-235. However, and this is very important, the nuclei of U-235 become much better catchers of neutrons that move very slowly. Thus if we could slow down the fission neutrons in such a way that their originally high velocity will be considerably reduced *before* they encounter on their way the first nucleus of uranium (238 or 235), the nuclei of U-235, though being in the minority, will have a better chance for capturing the neutrons than the nuclei of U-238.

The necessary slowing down arrangement can be achieved by distributing a large number of small pieces of natural uranium through some material (*moderator*) which slows down the neutrons without capturing too many of them. The best materials to be used for this purpose are heavy water, carbon, and the salts of beryllium. In Figure 76 we give a schematic picture of how such a "pile" formed by uranium grains distributed through a moderating substance actually works.[20]

As stated above, the light isotope, U-235 (which represents only 0.7 per cent of the natural uranium), is the only existing kind of fissionable nuclei capable of supporting a progressive chain reaction, thus leading to the large-scale liberation of nuclear energy. It does not mean, however, that we cannot build *artificially* other nuclear species, ordinarily not existing in nature, that would

[20] For more detailed discussion of uranium piles the reader is referred again to special books on atomic energy.

have the same properties as U-235. In fact, by using neutrons that are produced in large quantities by the progressive chain reaction in one fissionable element, we can turn other ordinarily unfissionable nuclei into fissionable ones.

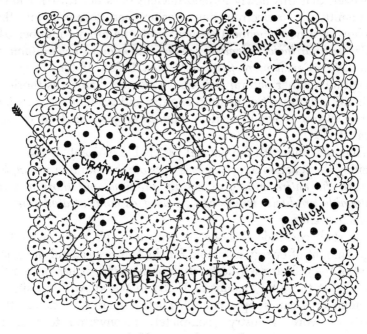

FIGURE 76

This somewhat biological looking picture represents the lumps of uranium (large atoms) imbedded in a moderator substance (small atoms). Two neutrons resulting from the fission of a U-nucleus in the lump on the left enter the moderator and are gradually slowed down by a series of collisions with its atomic nuclei. By the time these neutrons reach other uranium lumps they are considerably slowed down and are captured by U-235 nuclei which are much more efficient in respect to slow neutrons than are the nuclei of U-238.

The first example of this kind is demonstrated by the events taking place in the above described "pile," which uses natural uranium mixed with the moderating substance. We have seen that using the moderator we can reduce the neutron capture of U-238 nuclei, to the extent of permitting the development of a chain reaction between U-235 nuclei. However, some of the neutrons will still be captured by U-238. Where does this lead to?

The immediate result of neutron capture in U-238 is, of course, the still heavier uranium isotope, U-239. It was found, however, that this newly formed nucleus does not exist for long, and emitting two electrons one after another, goes over into the nucleus of a new chemical element with the atomic number 94. *This new artificial element, which is known as plutonium* (Pu-239), *is even more fissionable than U-235.* If for U-238 we substitute another natural radioactive element known as thorium (Th-232), the result of neutron capture and subsequent emission of two electrons would lead to *another artificial fissionable element*, U-233.

Thus, starting with the natural fissionable element, U-235, and running the reaction in cycles, *it is possible, certainly in principle, to turn the entire supply of natural uranium and thorium into fissionable products, which can be used as concentrated sources of nuclear energy.*

We shall conclude this section with a rough estimate of the total amount of energy available for the future peaceful development or the military self-destruction of mankind. It has been estimated that the total amount of U-235 in the known deposits of uranium ores can supply enough nuclear energy to satisfy the needs of world industry (completely reconverted to nuclear energy) for a period of several years. If, however, we take into account the possibility of using U-238 by turning it into plutonium, the time estimate will extend to several centuries. Throwing in the deposits of thorium (turned into U-233), which is about four times as abundant as uranium, we bring our estimate further up to at least one or two thousand years, which is long enough to make all worry about the "future shortages of atomic energy" unnecessary.

However, even if all these resources of nuclear energy are used, and no new deposits of uranium and thorium ores are discovered, future generations will still always be able to obtain nuclear energy from ordinary rocks. In fact, uranium and thorium, like all other chemical elements, are contained in small quantities in practically any ordinary material. Thus ordinary granite rocks contain 4 g of uranium and 12 g of thorium per ton. At first glance it looks like very little, but let us perform the following calculation. We know that one kilogram of fissionable material contains an amount of nuclear energy equivalent to 20,000 tons of TNT if it is exploded (as in an atomic bomb), or about 20,000 tons of gasoline if it is used as a fuel. Thus the 16 g of uranium and thorium contained in one ton of granite rock, if turned into fissionable materials, would be equivalent to 320 tons of ordinary fuel. That is enough to repay us for all the complicated trouble of separation—especially if we found that we were nearing the end of our supply of the richer deposits of ores.

Having conquered the energy liberation in nuclear fission of heavy elements such as uranium, physicists tackled the reverse process called *nuclear fusion*, in which two nuclei of light elements fuse together to form a heavier nucleus, liberating huge amounts of energy. As we shall see in Chapter XI, our sun gets its energy by such a fusion process, in which ordinary hydrogen nuclei unite to form the heavier nuclei of helium, as a result of violent thermal collisions in their interiors. To duplicate these so-called thermonuclear reactions for human purposes, the best material for producing fusion is heavy hydrogen, or *deuterium*, which is present in small amounts in ordinary water. The deuterium nucleus, called a deuteron, contains one proton and one neutron. When two deuterons collide, one of the following two reactions occurs:

2 deuterons → He-3 + neutron; 2 deuterons → H-3 + proton

In order to achieve the transformation, deuterium must be subjected to a temperature of a hundred million degrees.

The first successful nuclear-fusion device was the hydrogen bomb, in which the deuterium reaction was triggered by explosion of a fission bomb. A much more complex problem, however, is the production of *controlled thermonuclear reaction,* which would supply vast amounts of energy for peaceful purposes. The main difficulty—that of confining tremendously hot gas—can be overcome by means of strong magnetic fields that prevent the deuterons from touching the container's walls (which would melt and evaporate!) by confining them within a central hot region.

CHAPTER VIII

The Law of Disorder

1. THERMAL DISORDER

IF YOU pour a glass of water and look at it, you will see a clear uniform fluid with no trace of any internal structure or motion in it whatsoever (provided, of course, you do not shake the glass). We know, however, that the uniformity of water is only apparent and that if the water is magnified a few million times, there will be revealed a strongly expressed granular structure formed by a large number of separate molecules closely packed together.

Under the same magnification it is also apparent that the water is far from still, and that its molecules are in a state of violent agitation moving around and pushing one another as though they were people in a highly excited crowd. This irregular motion of water molecules, or the molecules of any other material substance, is known as *heat (or thermal) motion,* for the simple reason that it is responsible for the phenomenon of heat. For, although molecular motion as well as molecules themselves are not directly discernible to the human eye, it is molecular motion that produces a certain irritation in the nervous fibers of the human organism and produces the sensation that we call heat. For those organisms that are much smaller than human beings, such as, for example, small bacteria suspended in a water drop, the effect of thermal motion is much more pronounced, and these poor creatures are incessantly kicked, pushed, and tossed around by the restless molecules that attack them from all sides and give them no rest (Figure 77). This amusing phenomenon, known as *Brownian motion,* named after the English botanist Robert Brown, who first noticed it more than a century ago in a study of tiny plant spores, is of quite general nature and can be observed in the study of any kind of sufficiently small particles suspended in any kind of liquid, or of microscopic particles of smoke and dust floating in the air.

If we heat the liquid the wild dance of tiny particles suspended in it becomes more violent; with cooling the intensity of the motion noticeably subsides. This leaves no doubt that we are actually watching here the effect of the hidden thermal motion of matter, and that what we usually call temperature is nothing else but a measurement of the degree of molecular agitation. By studying the dependence of Brownian motion on temperature, it was found that at the temperature of $-273°$ C or $-459°$ F,

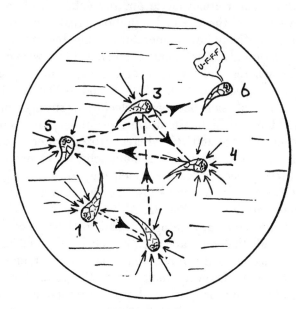

FIGURE 77

Six consecutive positions of a bacterium which is being tossed around by molecular impacts (physically correct; bacteriologically not quite so).

thermal agitation of matter completely ceases, and all its molecules come to rest. This apparently is the lowest temperature and it has received the name of *absolute zero*. It would be an absurdity to speak about still lower temperatures since apparently there is no motion slower than absolute rest!

Near the absolute zero temperature the molecules of any substance have so little energy that the cohesive forces acting upon them cement them together into one solid block, and all they can do is only quiver slightly in their frozen state. When the temperature rises the quivering becomes more and more intense, and at a certain stage our molecules obtain some freedom of motion and are able to slide by one another. The rigidity of the frozen substance disappears, and it becomes a fluid. The temperature at which the melting process takes place depends on the strength of the cohesive forces acting upon the molecules. In some materials such as hydrogen, or a mixture of nitrogen and oxygen which form atmospheric air, the cohesion of molecules is very weak, and the thermal agitation breaks up the frozen state at comparatively low temperatures. Thus hydrogen exists in the frozen state only at temperatures below 14° abs (i.e., below −259° C), whereas solid oxygen and nitrogen melt at 55° abs and 64° abs, respectively (i.e. −218° C and −209° C). In other substances the cohesion between molecules is stronger and they remain solid up to higher temperatures: thus pure alcohol remains frozen up to −130° C, whereas frozen water (ice) melts only at 0° C. Other substances remain solid up to much higher temperatures; a piece of lead will melt only at +327° C, iron at +1535° C, and the rare metal known as osmium remains solid up to the temperature of +2700° C. Although in the solid state of matter the molecules are strongly bound to their places, it does not mean at all that they are not affected by thermal agitation. Indeed, according to the fundamental law of heat motion, the amount of energy in every molecule is the same for all substances, solid, liquid, or gaseous at a given temperature, and the difference lies only in the fact that whereas in some cases this energy suffices to tear off the molecules from their fixed positions and let them travel around, in other cases they can only quiver on the same spot as angry dogs restricted by short chains.

This thermal quivering or vibration of molecules forming a solid body can be easily observed in the X-ray photographs described in the previous chapter. We have seen indeed that, since taking a picture of molecules in a crystal lattice requires a considerable time, it is essential that they should not move away from their fixed positions during the exposure. But a constant quivering around the fixed position is not conducive to good photography, and results in a somewhat blurred picture. This

FIGURE 78

effect is shown in the molecular photograph which is reproduced in Plate I. To obtain sharper pictures one must cool the
crystals as much as possible. This is sometimes accomplished by
dipping them in liquid air. If, on the other hand, one warms up
the crystal to be photographed, the picture becomes more and
more blurred, and, at the melting point the pattern completely
vanishes, owing to the fact that the molecules leave their places
and begin to move in an irregular way through the melted
substance.

After solid material melts, the molecules still remain together,
since the thermal agitation, though strong enough to dislocate
them from the fixed position in the crystalline lattice, is not yet
sufficient to take them completely apart. At still higher temperatures, however, the cohesive forces are not able to hold the
molecules together any more and they fly apart in all directions
unless prevented from doing so by the surrounding walls. When
this happens, of course, the result is matter in a gaseous state.
As in the melting of a solid, the evaporation of liquids takes place
at different temperatures for different materials, and the substances with a weaker internal cohesion will turn into vapor at
lower temperatures than those in which cohesive forces are
stronger. In this case the process also depends rather essentially
on the pressure under which the liquid is kept, since the outside
pressure evidently helps the cohesive forces to keep the molecules
together. Thus, as everybody knows, water in a tightly closed
kettle boils at a lower temperature than will water in an open one.
On the other hand, on the top of high mountains, where atmospheric pressure is considerably less, water will boil well below
100° C. It may be mentioned here that by measuring the temperature at which water will boil, one can calculate atmospheric
pressure and consequently the distance above sea level of a given
location.

But do not follow the example of Mark Twain who, according
to his story, once decided to put an aneroid barometer into a
boiling kettle of pea soup. This will not give you any idea of the
elevation, and the copper oxide will make the soup taste bad.

The higher the melting point of a substance, the higher is its
boiling point. Thus liquid hydrogen boils at −253° C, liquid

oxygen and nitrogen at $-183°$ C and $-196°$ C, alcohol at $+78°$ C, lead at $+1620°$ C, iron at $+3000°$ C and osmium only above $+5300°$ C.[1]

The breaking up of the beautiful crystalline structure of solid bodies forces the molecules first to crawl around one another like a pack of worms, and then to fly apart as though they were a flock of frightened birds. But this latter phenomenon still does not represent the limit of the destructive power of increasing thermal motion. If the temperature rises still farther the very existence of the molecules is threatened, since the ever increasing violence of intermolecular collisions is capable of breaking them up into separate atoms. This *thermal dissociation*, as it is called, depends on the relative strength of the molecules subjected to it. The molecules of some organic substances will break up into separate atoms or atomic groups at temperatures as low as a few hundred degrees. Other more sturdily built molecules, such as those of water, will require a temperature of over a thousand degrees to be destroyed. But when the temperature rises to several thousand degrees no molecules will be left and the matter will be a gaseous mixture of pure chemical elements.

This is the situation on the surface of our sun where the temperature ranges up to $6000°$ C. On the other hand, in the comparatively cooler atmospheres of the red stars,[2] some of the molecules are still present, a fact that has been demonstrated by the methods of spectral analysis.

The violence of thermal collisions at high temperatures not only breaks up the molecules into their constituent atoms, but also damages the atoms themselves by chipping off their outer electrons. This *thermal ionization* becomes more and more pronounced when the temperature rises into tens and hundreds of thousands of degrees, and reaches completion at a few million degrees above zero. At these tremendously hot temperatures, which are high above everything that we can produce in our laboratories but which are common in the interiors of stars and in particular inside our sun, the atoms as such cease to exist. All electronic shells are completely stripped off, and the matter

---

[1] All values given for atmospheric pressure.
See Chapter XI.

becomes a mixture of bare nuclei and free electrons rushing wildly through space and colliding with one another with tremendous force. However, in spite of the complete wreckage of atomic bodies, the matter still retains its fundamental chemical

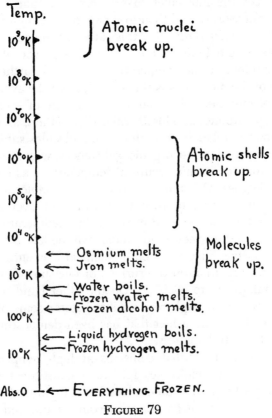

FIGURE 79

The destructive effect of temperature.

characteristics, inasmuch as atomic nuclei remain intact. If the temperature drops, the nuclei will recapture their electrons and the integrity of atoms will be reestablished.

In order to attain complete thermal dissociation of matter, that is to break up the nuclei themselves into the separate nucleons (protons and neutrons) the temperature must go up to at least several billion degrees. Even inside the hottest stars we do not find such high temperatures, though it seems very likely that temperatures of that magnitude did exist several billion years ago when our universe was still young. We shall return to this exciting question in the last chapter of this book.

Thus we see that the effect of thermal agitation is to destroy step by step the elaborate architecture of matter based on the law of quantum, and to turn this magnificent building into a mess of widely moving particles rushing around and colliding with one another without any apparent law or regularity.

2. HOW CAN ONE DESCRIBE DISORDERLY MOTION?

It would be, however, a grave mistake to think that because of the irregularity of thermal motion it must remain outside the scope of any possible physical description. Indeed the fact itself that thermal motion is *completely irregular* makes it subject to a new kind of law, the *Law of Disorder* better known as the *Law of Statistical Behavior*. In order to understand the above statement let us turn our attention to the famous problem of a "Drunkard's Walk." Suppose we watch a drunkard who has been leaning against a lamp post in the middle of a large paved city square (nobody knows how or when he got there) and then has suddenly decided to go nowhere in particular. Thus off he goes, making a few steps in one direction, then some more steps in another, and so on and so on, changing his course every few steps in an entirely unpredictable way (Figure 80). How far will be our drunkard from the lamp post after he has executed, say, a hundred phases of his irregular zigzag journey? One would at first think that, because of the unpredictability of each turn, there is no way of answering this question. If, however, we consider the problem a little more attentively we will find that, although we really cannot tell where the drunkard will be at the end of his walk, we can answer the question about his *most probable* distance from the lamp post after a given large number of turns. In order to approach this problem in a vigorous mathematical way let us draw on the pavement two co-ordinate axes with the origin in the lamp post; the X-axis coming toward us and the Y-axis to the right. Let R be the distance of the drunkard from the lamp post after the total of N zigzags (14 in Figure 80). If now X_N and Y_N are the projections of the Nth leg of the track on the corresponding axis, the Pythagorean theorem gives us apparently:

$$R^2 = (X_1 + X_2 + X_3 \cdots + X_N)^2 + (Y_1 + Y_2 + Y_3 + \cdots Y_N)^2$$

where X's and Y's are positive or negative depending on whether our drunkard was moving to or from the post in this particular

FIGURE 80
Drunkard's walk.

phase of his walk. Notice that since his motion is *completely disorderly*, there will be about as many positive values of X's and Y's as there are negative. In calculating the value of the square of the terms in parentheses according to the elementary rules of algebra, we have to multiply each term in the bracket by itself and by each of all other terms.

Thus:

$$(X_1+X_2+X_3+\cdots X_N)^2$$
$$=(X_1+X_2+X_3+\cdots X_N)\ (X_1+X_2+X_3+\cdots X_N)$$
$$=X_1{}^2+X_1X_2+X_1X_3+\cdots X_2{}^2+X_1X_2+\cdots X_N{}^2$$

This long sum will contain the square of all X's ($X_1{}^2$, $X_2{}^2\cdots X_N{}^2$), and the so-called "mixed products" like X_1X_2, X_2X_3, etc.

So far it is simple arithmetic, but now comes the statistical point based on the disorderliness of the drunkard's walk. Since he was moving entirely at random and would just as likely make a step toward the post as away from it, the values of X's have a fifty-fifty chance of being either positive or negative. Consequently in looking through the "mixed products" you are likely to find always the pairs that have the same numerical value but opposite signs thus canceling each other, and the larger the total number of turns, the more likely it is that such a compensation takes place. What will be left are only the squares of X's, since the square is always positive. Thus the whole thing can be written as $X_1{}^2+X_2{}^2+\cdots X_N{}^2=N\ X^2$ where X is the average length of the projection of a zigzag link on the X-axis.

In the same way we find that the second bracket containing Y's can be reduced to: NY^2, Y being the average projection of the link on the Y-axis. It must be again repeated here that what we have just done is not strictly an algebraic operation, but is based on the statistical argument concerning the mutual cancellation of "mixed products" because of the random nature of the pass. For the most probable distance of our drunkard from the lamp post we get now simply:

$$R^2=N\ (X^2+Y^2)$$

or

$$R=\sqrt{N}\cdot\sqrt{X^2+Y^2}$$

But the average projections of the link on both axes is simply a 45° projection, so that $\sqrt{X^2+Y^2}$ is (again because of the Pythagorean theorem) simply equal to the average length of the link. Denoting it by 1 we get:

$$R=1\cdot\sqrt{N}$$

In plain words our result means: *the most probable distance of our drunkard from the lamp post after a certain large number of irregular turns is equal to the average length of each straight track that he walks, times the square root of their number.*

Thus if our drunkard goes one yard each time before he turns (at an unpredictable angle!), he will most probably be only ten yards from the lamp post after walking a grand total of a hundred yards. If he had not turned, but had gone straight, he would be a hundred yards away—which shows that it is definitely advantageous to be sober when taking a walk.

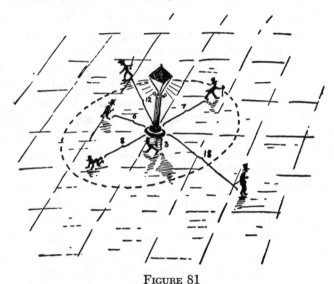

FIGURE 81

Statistical distribution of six walking drunkards around the lamp post.

The statistical nature of the above example is revealed by the fact that we refer here only to the *most probable* distance and not to the exact distance in each individual case. In the case of an individual drunkard it may happen, though this is not very probable, that he does not make any turns at all and thus goes far away from the lamp post along the straight line. It may also happen, that he turns each time by, say, 180 degrees thus returning to the lamp post after every second turn. But if a large number of drunkards all start from the same lamp post walking in different zigzag paths and not interfering with one another you will find after a sufficiently long time that they are spread over a certain area around the lamp post in such a way that their *average distance* from the post may be calculated by the above rule. An example of such spreading due to irregular motion is given in Figure 81, where we consider six walking drunkards. It goes without saying that the larger the number of drunkards, and the larger the number of turns they make in their disorderly walk, the more accurate is the rule.

Now substitute for the drunkards some microscopic bodies such as plant spores or bacteria suspended in liquid, and you will have exactly the picture that the botanist Brown saw in his microscope. True the spores and bacteria are not drunk, but, as we have said above, they are being incessantly kicked in all possible directions by the surrounding molecules involved in thermal motion, and are therefore forced to follow exactly the same irregular zigzag trajectories as a person who has completely lost his sense of direction under the influence of alcohol.

If you look through a microscope at the Brownian motion of a large number of small particles suspended in a drop of water, you will concentrate your attention on a certain group of them that are at the moment concentrated in a given small region (near the "lamp post"). You will notice that in the course of time they become gradually dispersed all over the field of vision, and that their average distance from the origin increases in proportion to the square root of the time interval as required by the mathematical law by which we calculated the distance of the drunkard's walk.

The same law of motion pertains, of course, to each separate molecule in our drop of water; but you cannot see separate molecules, and even if you could, you wouldn't be able to distinguish between them. To make such motion visible one must use two different kinds of molecules distinguishable for example by their different colors. Thus we can fill one half of a chemical test tube with a water solution of potassium permanganate, which will give to the water a beautiful purple tint. If we now pour on the top of it some clear fresh water, being careful not to mix up the two layers, we shall notice that the color gradually penetrates the clear water. If you wait sufficiently long you will find that all the water from the bottom to the surface becomes uniformly colored. This phenomenon, familiar to everybody, is known as *diffusion* and is due to the irregular thermal motion of the molecules of dye among the water molecules. We must imagine each molecule of potassium permanganate as a little drunkard who is driven to and fro by the incessant impacts received from other molecules. Since in water the molecules are packed rather tightly (in contrast to the arrangement of those in a gas) the average free path of each molecule between two successive collisions is very short, being only about one hundred millionths of an inch. Since on the other hand the molecules at room temperature move with the speed of about one tenth of a mile per second, it takes only one million-millionth part of a second for a molecule to go from one collision to another. Thus in the course of a single second

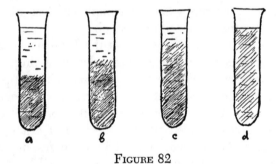

FIGURE 82

each dye molecule will be engaged in about a million million consecutive collisions and will change its direction of motion as many times. The average distance covered during the first second will be one hundred millionth of an inch (the length of free path) times the square root of a million millions. This gives the average diffusion speed of only one hundredth of an inch per second; a rather slow progress considering that if it were not deflected by collisions, the same molecule would be a tenth of a mile away! If you wait 100 sec, the molecule will have struggled through 10 times ($\sqrt{100}=10$) as great distance, and in 10,000 sec, that is, in about 3 hr, the diffusion will have carried the coloring 100 times farther ($\sqrt{10000}=100$), that is, about 1 in. away. Yes, diffusion is a rather slow process; when you put a lump of sugar into your cup of tea you had better stir it rather than wait until the sugar molecules have been spread throughout by their own motion.

Just to give another example of the process of diffusion, which is one of the most important processes in molecular physics, let us consider the way in which heat is propagated through an iron poker, one end of which you put into the fireplace. From your own experience you know that it takes quite a long time until the other end of the poker becomes uncomfortably hot, but you probably do not know that the heat is carried along the metal stick by the process of diffusion of electrons. Yes, an ordinary iron poker is actually stuffed with electrons, and so is any metallic object. The difference between a metal, and other materials, as for example glass, is that the atoms of the former lose some of their outer electrons, which roam all through the metallic lattice, being involved in irregular thermal motion, in very much the same way as the particles of ordinary gas.

The surface forces on the outer boundaries of a piece of metal prevent these electrons from getting out,[3] but in their motion inside the material they are almost perfectly free. If an electric force is applied to a metal wire, the free unattached electrons will rush headlong in the direction of the force producing the phenomenon of electric current. The nonmetals on the other hand are usually good insulators because all their electrons are bound to atoms and thus cannot move freely.

When one end of a metal bar is placed in the fire, the thermal motion of free electrons in this part of the metal is considerably increased, and the fast-moving electrons begin to diffuse into the other regions carrying with them the extra energy of heat. The process is quite similar to the diffusion of dye molecules through water, except that instead of having two different kinds of particles (water molecules and dye molecules) we have here the *diffusion of hot electron gas into the region occupied by cold electron gas*. The drunkard's walk law applies here, however, just

---

[3] When we bring a metal wire to a high temperature, the thermal motion of electrons in its inside becomes more violent and some of them come out through the surface. This is the phenomenon used in electron tubes and familiar to all radio amateurs.

as well and the distances through which the heat propagates along a metal bar increase as the square roots of corresponding times.

As our last example of diffusion we shall take an entirely different case of cosmic importance. As we shall learn in the following chapters the energy of our sun is produced deep in its interior by the alchemic transformation of chemical elements. This energy is liberated in the form of intensive radiation, and the "particles of light," or the light quanta begin their long journey through the body of the sun towards its surface. Since light moves at a speed of 300,000 km per second, and the radius of the sun is only 700,000 km it would take a light quantum only slightly over two seconds to come out provided it moved without any deviations from a straight line. However, this is far from being the case; on their way out the light quanta undergo innumerable collisions with the atoms and electrons in the material of the sun. The free pass of a light quantum in solar matter is about a centimeter (much longer than a free pass of a molecule!) and since the radius of the sun is 70,000,000,000 cm, our light quantum must make $(7\cdot10^{10})^2$ or $5\cdot10^{21}$ drunkard's steps to reach the surface. Since each step requires $\dfrac{1}{3\cdot10^{10}}$ or $3\cdot10^{-11}$ sec, the entire time of travel is $3\times10^{-11}\times5\times10^{21}=1.5\times10^{11}$ sec or about 5000 yr! Here again we see how slow the process of diffusion is. It takes light 50 centuries to travel from the center of the sun to its surface, whereas after coming into empty interplanetary space and traveling along a straight line it covers the entire distance from the sun to the earth in only eight minutes!

3. *COUNTING PROBABILITIES*

This case of diffusion represents only one simple example of the application of the statistical law of probability to the problem of molecular motion. Before we go farther with that discussion, and make the attempt to understand the all-important *Law of Entropy*, which rules the thermal behavior of every material body, be it a tiny droplet of some liquid or the giant universe of stars, we have first to learn more about the ways in which the probability of different simple or complicated events can be calculated.

By far the simplest problem of probability calculus arises when you toss a coin. Everybody knows that in this case (without cheating) there are equal chances to get heads or tails. One usually says that there is a *fifty-fifty chance* for heads or tails, but it is more customary in mathematics to say that the chances are *half and half*. If you add the chances of getting heads and getting tails you get $\frac{1}{2}+\frac{1}{2}=1$. Unity in the theory of probability means a certainty; you are in fact quite certain that in tossing a

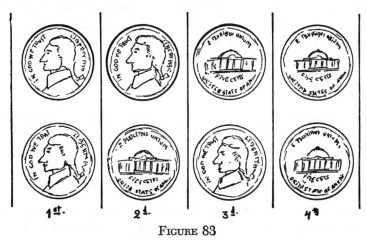

FIGURE 83

Four possible combinations in tossing two coins.

coin you get either heads or tails, unless it rolls under the sofa and vanishes tracelessly.

Suppose now you drop the coin twice in succession or, what is the same, you drop 2 coins simultaneously. It is easy to see that you have here 4 different possibilities shown in Figure 83.

In the first case you get heads twice, in the last case tails twice, whereas the two intermediate cases lead to the same result since it does not matter to you in which order (or in which coin) heads or tails appear. Thus you say that the chances of getting heads twice are 1 out of 4 or $\frac{1}{4}$, the chances of getting tails twice are also $\frac{1}{4}$, whereas the chances of heads once and tails once are 2 out of 4 or $\frac{1}{2}$. Here again $\frac{1}{4}+\frac{1}{4}+\frac{1}{2}=1$, meaning that you are certain to get one of the 3 possible combinations. Let us see now what happens if we toss the coin 3 times. There are altogether 8 possibilities summarized in the following table:

First tossing	h	h	h	h	t	t	t	t
Second	h	h	t	t	h	h	t	t
Third	h	t	h	t	h	t	h	t
	I	II	II	III	II	III	III	IV

If you inspect this table you find that there is 1 chance out of 8 of getting heads three times, and the same of getting tails three times. The remaining possibilities are equally divided between heads twice and tails once, or heads once and tails twice, with the probability three eighths for each event.

Our table of different possibilities is growing rather rapidly, but let us take one more step by tossing 4 times. Now we have the following 16 possibilities:

First tossing	h	h	h	h	h	h	h	h	t	t	t	t	t	t	t	t
Second	h	h	h	h	t	t	t	t	h	h	h	h	t	t	t	t
Third	h	h	t	t	h	h	t	t	h	h	t	t	h	h	t	t
Fourth	h	t	h	t	h	t	h	t	h	t	h	t	h	t	h	t
	I	II	II	III	II	III	III	IV	II	III	III	IV	III	IV	IV	V

Here we have $\frac{1}{16}$ for the probability of heads four times, and exactly the same for tails four times. The mixed cases of heads three times and tails once or tails three times and heads once have the probabilities of $\frac{4}{16}$ or $\frac{1}{4}$ each, whereas the chances of heads and tails the same number of times are $\frac{6}{16}$ or $\frac{3}{8}$.

If you try to continue in a similar way for larger numbers of tosses the table becomes so long that you will soon run out of paper; thus for example for ten tosses you have 1024 different possibilities (i.e., $2 \times 2 \times 2 \times 2 \times 2 \times 2 \times 2 \times 2 \times 2 \times 2$). But it is not at all necessary to construct such long tables since the simple laws of probability can be observed in those simple examples that we already have cited and then used directly in more complicated cases.

First of all you see that the probability of getting heads twice is equal to the product of the probabilities of getting it separately in the first and in the second tossing; in fact $\frac{1}{4} = \frac{1}{2} \times \frac{1}{2}$. Similarly the probability of getting heads three or four times in succession is the product of probabilities of getting it separately in each tossing ($\frac{1}{8}=\frac{1}{2}\times\frac{1}{2}\times\frac{1}{2}$; $\frac{1}{16}=\frac{1}{2}\times\frac{1}{2}\times\frac{1}{2}\times\frac{1}{2}$). Thus if somebody asks you what the chances are of getting heads each time in ten tossings you can easily give the answer by multiplying $\frac{1}{2}$ by $\frac{1}{2}$ ten times. The result will be .00098, indicating that the chances are very low indeed: about one chance out of a thousand! Here we have the rule of "multiplication of probabilities," which states that *if you want several different things, you may determine the mathematical probability of getting them by multiplying the mathematical probabilities of getting the several individual ones.* If there are many things you want, and each of them is not particularly probable, the chances that you get them *all* are discouragingly low!

There is also another rule, that of the "addition of probabilities," which states that *if you want only one of several things (no matter which one), the mathematical probability of getting it is the sum of mathematical probabilities of getting individual items on your list.*

This can be easily illustrated in the example of getting an equal division between heads and tails in tossing a coin twice. What you actually want here is *either* "heads once, tails twice" or "tails twice, heads once." The probability of each of the above combinations is $\frac{1}{4}$, and the probability of getting either one of them is $\frac{1}{4}$ plus $\frac{1}{4}$ or $\frac{1}{2}$. Thus: If you want "that, *and* that, *and* that . . ." you *multiply* the individual mathematical probabilities of different items. If, however, you want "that, *or* that, *or* that" you *add* the probabilities.

In the first case your chances of getting everything you ask for will decrease as the number of desired items increases. In the second case, when you want only one out of several items your chances of being satisfied increase as the list of items from which to choose becomes longer.

The experiments with tossing coins furnish a fine example of what is meant by saying that the laws of probability become more exact when you deal with a large number of trials. This is illustrated in Figure 84, which represents the probabilities of getting a different relative number of heads and tails for two, three, four, ten, and a hundred tossings. You see that with the increasing number of tossings the probability curve becomes sharper and sharper and the maximum at fifty-fifty ratio of heads and tails becomes more and more pronounced.

Thus whereas for 2 or 3, or even 4 tosses, the chances to have heads each time or tails each time are still quite appreciable, in 10 tosses even 90 per cent of heads or tails is very improbable.

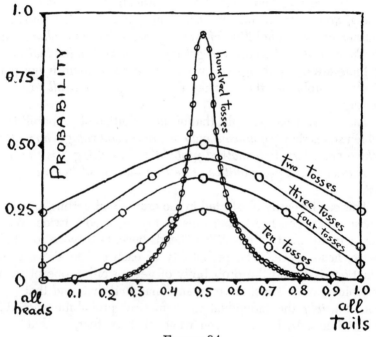

FIGURE 84

Relative number of tails and heads.

For a still larger number of tosses, say 100 or 1000, the probability curve becomes as sharp as a needle, and the chances of getting even a small deviation from fifty-fifty distribution becomes practically nil.

Let us now use the simple rules of probability calculus that we have just learned in order to judge the relative probabilities of various combinations of five playing cards which one encounters in the well-known game of poker.

In case you do not know, each player in this game is dealt 5 cards and the one who gets the highest combination takes the bank. We shall omit here the additional complications arising from the possibility of exchanging some of your cards with the hope of getting better ones, and the psychological strategy of bluffing your opponents into submission by making them believe that you have much better cards than you actually have. Although this bluffing actually is the heart of the game, and once led the famous Danish physicist Niels Bohr to propose an entirely new type of game in which no cards are used, and the players simply bluff one another by talking about the imaginary combinations they have, it lies entirely outside the domain of probability calculus, being a purely psychological matter.

FIGURE 85

A flush (of spades).

In order to get some exercise in probability calculus, let us calculate the probabilities of some of the combinations in the game of poker. One of these combinations is called a "flush" and represents 5 cards all of the same suit (Figure 85).

If you want to get a flush it is immaterial what the first card you get is, and one has only to calculate the chances that the other four will be of the same suit. There are altogether 52 cards in the pack, 13 cards of each suit,[4] so that after you get your first card, there remain in the pack 12 cards of the same suit. Thus the chances that your second card will be of the proper suit are 12/51. Similarly the chances that the third, fourth, and fifth cards

[4] We omit here the complications arising from the presence of the "joker," an extra card which can be substituted for any other card according to the desire of the player.

will be of the same suit are given by the fractions: 11/50, 10/49 and 9/48. Since you want *all* 5 cards to be of the same suit you have to apply the rule of probability-multiplications. Doing this you find that the probability of getting a flush is:

$$\frac{12}{51} \times \frac{11}{50} \times \frac{10}{49} \times \frac{9}{48} = \frac{13068}{5997600} \text{ or about 1 in 500.}$$

But please do not think that in 500 hands you are sure to get a flush. You may get none, or you may get two. This is only probability calculus, and it may happen that you will be dealt many more than 500 hands without getting the desired combination, or on the contrary that you may be dealt a flush the very first time you have the cards in your hands. All that the theory of prob-

Figure 86
Full house.

ability can tell you is that you will *probably* be dealt 1 flush in 500 hands. You may also learn, by following the same methods of calculation, that in playing 30,000,000 games you will probably get 5 aces (including the joker) about ten times.

Another combination in poker, which is even rarer and therefore more valuable, is the so-called "full hand," more popularly called "full house." A full house consists of a "pair" and "three of a kind" (that is, 2 cards of the same value in 2 suits, and 3 cards of the same value in 3 suits—as, for example, the 2 fives and 3 queens shown in Figure 86).

If you want to get a full house, it is immaterial which 2 cards you get first, but when you get them you must have 2 of the remaining 3 cards match one of them, and the other match the other one. Since there are 6 cards that will match the ones you have (if you have a queen and a five, there are 3 other queens and 3 other fives) the chances that the third card is a right one are 6 out of 50 or 6/50. The chances that the fourth card will be the right one are 5/49 since there are now only 5 right cards out of 49 cards left, and the chance that the fifth card will be right is 4/48. Thus the total probability of a full house is:

$$\frac{6}{50} \times \frac{5}{49} \times \frac{4}{48} = \frac{120}{117600}$$

or about one half of the probability of the flush.

In a similar way one can calculate the probabilities of other combinations as, for example, a "straight" (a sequence of cards), and also take into account the changes in probability introduced by the presence of the joker and the possibility of exchanging the originally dealt cards.

By such calculations one finds that the sequence of seniority used in poker does really correspond to the order of mathematical probabilities. It is not known by the author whether such an arrangement was proposed by some mathematician of the old times, or was established purely empirically by millions of players risking their money in fashionable gambling salons and little dark haunts all over the world. If the latter was the case, we must admit that we have here a pretty good statistical study of the relative probabilities of complicated events!

Another interesting example of probability calculation, an example that leads to a quite unexpected answer, is the problem of "Coinciding Birthdays." Try to remember whether you have ever been invited to two different birthday parties on the same day. You will probably say that the chances of such double invitations are very small since you have only about 24 friends who are likely to invite you, and there are 365 days in the year on which their birthdays may fall. Thus, with so many possible dates to choose from, there must be very little chance that any 2 of your 24 friends will have to cut their birthday cakes on the same day.

However, unbelievable as it may sound, your judgment here is quite wrong. The truth is that there is a rather high probability that in a company of 24 people there are a pair, or even several pairs, with coinciding birthdays. As a matter of fact, there are more chances that there is such a coincidence than that there is not.

You can verify that fact by making a birthday list including about 24 persons, or more simply, by comparing the birth dates of 24 persons whose names appear consecutively on any pages of some such reference book as "Who's Who in America," opened at random. Or the probabilities can be ascertained by using the simple rules of probability calculus with which we have become acquainted in the problems of coin tossing and poker.

Suppose we try first to calculate the chances that in a company of twenty-four persons everyone has a different birth date. Let us ask the first person in the group what is his birth date; of course this can be any of the 365 days of the year. Now, what is the chance that the birth date of the second person we approach is *different* from that of the first? Since this (second) person could have been born on any day of the year, there is one chance out of 365 that his birth date coincides with that of the first one, and 364 chances out of 365 (i.e., the probability of 364/365) that it does not. Similarly, the probability that the third person has a birth date different from that of either the first or second is 363/365, since two days of the year have been excluded. The probabilities that the next persons we ask have different birth dates from the ones we have approached before are then: 362/365, 361/365, 360/365 and so on up to the last person for whom the probability is $\dfrac{(365-23)}{365}$ or $\dfrac{342}{365}$.

Since we are trying to learn what the probability is that one of these coincidences of birth dates exists, we have to multiply all the above fractions, thus obtaining for the probability of all the persons having different birth dates the value:

$$\frac{364}{365} \times \frac{363}{365} \times \frac{362}{365} \times \cdots \frac{342}{365}$$

One can arrive at the product in a few minutes by using certain methods of higher mathematics, but if you don't know them you can do it the hard way by direct multiplication,[5] which would not take so very much time. The result is 0.46, indicating

[5] Use a logarithmic table or slide rule if you can!

that the probability that there will be no coinciding birthdays is slightly less than one half. In other words there are only 46 chances in 100 that no two of your two dozen friends will have birthdays on the same day, and 54 chances in 100 that two or more will. Thus if you have 25 or more friends, and have never been invited to two birthday parties on the same date you may conclude with a high degree of probability that either most of your friends do not organize their birthday parties, or that they do not invite you to them!

The problem of coincident birthdays represents a very fine example of how a common-sense judgment concerning the probabilities of complex events can be entirely wrong. The author has put this question to a great many people, including many prominent scientists, and in all cases except one[6] was offered bets ranging from 2 to 1 to 15 to 1 that no such coincidence will occur. If he had accepted all these bets he would be a rich man by now!

It cannot be repeated too often that if we calculate the probabilities of different events according to the given rules and pick out the most probable of them, we are not at all sure that this is exactly what is going to happen. Unless the number of tests we are making runs into thousands, millions or still better into billions, the predicted results are only "likely" and not at all "certain." This slackening of the laws of probability when dealing with a comparatively small number of tests limits, for example, the usefulness of statistical analysis for deciphering various codes and cryptograms which are limited only to comparatively short notes. Let us examine, for example, the famous case described by Edgar Allan Poe in his well-known story "The Gold Bug." He tells us about a certain Mr. Legrand who, strolling along a deserted beach in South Carolina, picked up a piece of parchment half buried in the wet sand. When subjected to the warmth of the fire burning gaily in Mr. Legrand's beach hut, the parchment revealed some mysterious signs written in ink which was invisible when cold, but which turned red and was quite legible when heated. There was a picture of a skull, suggesting that the docu-

---

[6] This exception was, of course, a Hungarian mathematician (see the beginning of the first chapter of this book).

ment was written by a pirate, the head of a goat, proving beyond any doubt that the pirate was none other than the famous Captain Kidd, and several lines of typographical signs apparently indicating the whereabouts of a hidden treasure (see Figure 87).

We take it on the authority of Edgar Allan Poe that the pirates of the seventeenth century were acquainted with such typographical signs as semicolons and quotation marks, and such others as: ‡, +, and ¶.

Being in need of money, Mr. Legrand used all his mental powers in an attempt to decipher the mysterious cryptogram and

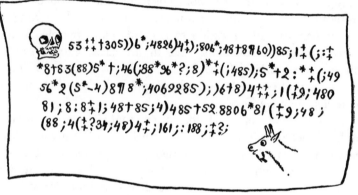

FIGURE 87

Captain Kidd's Message.

finally did so on the basis of the relative frequency of occurrence of different letters in the English language. His method was based on the fact that if you count the number of different letters of any English text, whether in a Shakespearian sonnet or an Edgar Wallace mystery story, you will find that the letter "e" occurs by far most frequently. After "e" the succession of most frequent letters is as follows:

a, o, i, d, h, n, r, s, t, u, y, c, f, g, l, m, w, b, k, p, q, x, z

By counting the different symbols appearing in Captain Kidd's cryptogram, Mr. Legrand found that the symbol that occurred most frequently in the message was the figure 8. "Aha," he said, "that means that 8 most probably stands for the letter e."

Well, he was right in this case, but of course it was only *very probable* and not at all certain. In fact if the secret message had been "You will find a lot of gold and coins in an iron box in woods two thousand yards south from an old hut on Bird Island's north tip" it would not have contained a single "e"! But the laws of chance were favorable to Mr. Legrand, and his guess was really correct.

Having met with success in the first step, Mr. Legrand became overconfident and proceeded in the same way by picking up the letters in the order of the probability of their occurrence. In the following table we give the symbols appearing in Captain Kidd's message in the order of their relative frequency of use:

Of the character 8 there are 33	e ← → e
;	26	a t
4	19	o h
‡	16	i o
(	16	d r
*	13	h n
5	12	n a
6	11	r i
†	8	s d
1	8	t
0	6	u
g	5	y
2	5	c
i	4	
3	4	g ← → g
?	3	l u
¶	2	m
-	1	w
.	1	b

The first column on the right contains the letters of the alphabet arranged in the order of their relative frequency in the

English language. Therefore it was logical to assume that the signs listed in the broad column to the left stood for the letters listed opposite them in the first narrow column to the right. But using this arrangement we find that the beginning of Captain Kidd's message reads: *ngiisgunddrhaoecr . . .*

No sense at all!

What happened? Was the old pirate so tricky as to use special words that do not contain letters that follow the same rules of frequency as those in the words normally used in the English language? Not at all; it is simply that the text of the message is not long enough for good statistical sampling and the most prob-able distribution of letters does not occur. Had Captain Kidd hidden his treasure in such an elaborate way that the instructions for its recovery occupied a couple of pages, or, still better an entire volume, Mr. Legrand would have had a much better chance to solve the riddle by applying the rules of frequency.

If you drop a coin 100 times you may be pretty sure that it will fall with the head up about 50 times, but in only 4 drops you may have heads three times and tails once or vice versa. To make a rule of it, the larger the number of trials, the more accurately the laws of probability operate.

Since the simple method of statistical analysis failed because of an insufficient number of letters in the cryptogram, Mr. Le-grand had to use an analysis based on the detailed structure of different words in the English language. First of all he strength-ened his hypothesis that the most frequent sign 8 stood for *e* by noticing that the combination 88 occurred very often (5 times) in this comparatively short message, for, as everyone knows, the letter e is very often doubled in English words (as in: *meet, fleet, speed, seen, been, agree, etc.*). Furthermore if 8 really stood for *e* one would expect it to occur very often as a part of the word "the." Inspecting the text of the cryptogram we find that the combination *;48* occurs seven times in a few short lines. But if this is true, we must conclude that *;* stands for *t* and *4* for *h*.

We refer the reader to the original Poe story for the details concerning the further steps in the deciphering of Captain Kidd's message, the complete text of which was finally found to be: "A good glass in the bishop's hostel in the devil's seat. Forty-one degrees and thirteen minutes northeast by north. Main branch seventh limb east side. Shoot from the left eye of the death's head. A bee-line from the tree through the shot fifty feet out."

The correct meaning of the different characters as finally deciphered by Mr. Legrand is shown in the second column of the table on page 217, and you see that they do not correspond exactly to the distribution that might reasonably be expected on the basis of the laws of probability. It is, of course, because the text is too short and therefore does not furnish an ample opportunity

FIGURE 88

for the laws of probability to operate. But even in this small "statistical sample" we can notice the tendency for the letters to arrange themselves in the order required by the theory of probability, a tendency that would become almost an unbreakable rule if the number of letters in the message were much larger.

There seems to be only one example (excepting the fact that insurance companies do not break up) in which the predictions of the theory of probability have actually been checked by a very large number of trials. This is a famous problem of the American flag and a box of kitchen matches.

To tackle this particular problem of probability you will need an American flag, that is, the part of it consisting of red and white stripes; if no flag is available just take a large piece of paper and draw on it a number of parallel and equidistant lines. Then you need a box of matches—any kind of matches, provided they are shorter than the width of the stripes. Next you will need a Greek pi, which is not something to eat, but just a letter of the Greek alphabet equivalent to our "p." It looks like this: π. In addition to being a letter of the Greek alphabet, it is used to

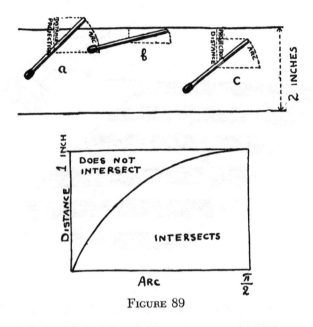

FIGURE 89

signify the ratio of the circumference of a circle to its diameter. You may know that numerically it equals 3.1415926535 . . . (many more digits are known, but we shall not need them all.)

Now spread the flag on a table, toss a match in the air and watch it fall on the flag (Figure 88). It may fall in such a way that it all remains within one stripe, or it may fall across the boundary between two stripes. What are the chances that one or another will take place?

Following our procedure in ascertaining other probabilities, we must first count the number of cases that correspond to one or another possibility.

But how can you count all the possibilities when it is clear that a match can fall on a flag in an infinite number of different ways?

Let us examine the question a little more closely. The position of the fallen match in respect to the stripe on which it falls can be characterized by the distance of the middle of the match from the nearest boundary line, and by the angle that the match forms with the direction of the stripes in Figure 89. We give three typical examples of fallen matches, assuming, for the sake of simplicity, that the length of the match equals the width of the stripe, each being, say, two inches. If the center of the match is rather close to the boundary line, and the angle is rather large (as in case *a*) the match will intersect the line. If, on the contrary, the angle is small (as in case *b*) or the distance is large (as in case *c*) the match will remain within the boundaries of one stripe. More exactly we may say that the match will intersect the line if the projection of the half-of-the-match on the vertical direction is larger than the half width of the stripe (as in case *a*), and that no intersection will take place if the opposite is true (as in case *b*). The above statement is represented graphically on the diagram in the lower part of the picture. We plot on the horizontal axis (abscissa) the angle of the fallen match as given by the length of the corresponding arc of radius 1. On the vertical axis (ordinate) we plot the length of the projection of the half-match length on the vertical direction; in trigonometry this length is known as the *sine* corresponding to the given arc. It is clear that the sine is zero when the arc is zero since in that case the match occupies a horizontal position. When the arc is ½ π, which corresponds to a right angle,[7] the sine is equal to unity, since the match occupies a vertical position and thus coincides with its projection. For intermediate values of the arc the sine is given by the familiar mathematical wavy curve known as sinusoid. (In Figure 89 we have only one quarter of a complete wave in the interval between 0 and $\pi/2$.)

[7] The circumference of a circle with the radius 1 is π times its diameter or 2π. Thus the length of one quadrant of a circle is $2\pi/4$ or $\pi/2$.

Having constructed this diagram we can use it with convenience for estimating the chances that the fallen match will or will not cross the line. In fact, as we have seen above (look again at the three examples in the upper part of Figure 89) the match will cross the boundary line of a stripe if the distance of the center of the match from the boundary line is less than the corresponding projection, that is, less than the sine of the arc. That means that in plotting that distance and that arc in our diagram we get a point *below* the sine line. On the contrary the match that falls entirely within the boundaries of a stripe will give a point *above* the sine line.

Thus, according to our rules for calculating probabilities, the chances of intersection will stand in the same ratio to the chances of nonintersection as the area below the curve does to the area above it; or the probabilities of the two events may be calculated by dividing the two areas by the entire area of the rectangle. It can be proved mathematically (*cf.* Chapter II) that the area of the sinusoid presented in our diagram equals exactly 1. Since the total area of the rectangle is $\frac{\pi}{2} \times 1 = \frac{\pi}{2}$ we find the probability that the match will fall across the boundary (for matches equal in length to the stripe width) is: $\frac{1}{\pi/2} = \frac{2}{\pi}$.

The interesting fact that π pops up here where it might be least expected was first observed by the eighteenth century scientist Count Buffon, and so the match-and-stripes problem now bears his name.

An actual experiment was carried out by a diligent Italian mathematician, Lazzerini, who made 3408 match tosses and observed that 2169 of them intersected the boundary line. The exact record of this experiment, checked with the Buffon formula, substitutes for π a value of $\frac{2+3408}{2169}$ or 3.1415929, differing from the exact mathematical value only in the seventh decimal place!

This represents, of course, a most amusing proof of the validity of the probability laws, but not more amusing than the determination of a number "2" by tossing a coin several thousand times and dividing the total number of tosses by the number of times heads come up. Sure enough you get in this case: 2.000000 . . . with just as small an error as in Lazzerini's determination of π.

4. THE "MYSTERIOUS" ENTROPY

From the above examples of probability calculus, all of them pertaining to ordinary life, we have learned that predictions of that sort, being often disappointing when small numbers are involved, become better and better when we go to really large numbers. This makes these laws particularly applicable to the description of the almost innumerable quantities of atoms or molecules that form even the smallest piece of matter we can conveniently handle. Thus, whereas the statistical law of Drunkard's Walk can give us only approximate results when applied to a half-dozen drunkards who make perhaps two dozen turns each, its application to billions of dye molecules undergoing billions of collisions every second leads to the most rigorous physical law of diffusion. We can also say that the dye that was originally dissolved in only one half of the water in the test tube tends through the process of diffusion to spread uniformly through the entire liquid, because, such uniform distribution is *more probable* than the original one.

For exactly the same reason the room in which you sit reading this book is filled uniformly by air from wall to wall and from floor to ceiling, and it never even occurs to you that the air in the room can unexpectedly collect itself in a far corner, leaving you to suffocate in your chair. However, *this horrifying event is not at all physically impossible, but only highly improbable.*

To clarify the situation, let us consider a room divided into two equal halves by an imaginary vertical plane, and ask ourselves about the most probable distribution of air molecules between the two parts. The problem is of course identical with the coin-tossing problem discussed in the previous chapter. If we pick up one single molecule it has equal chances of being in the right or in the left half of the room, in exactly the same way as the tossed coin can fall on the table with heads or tails up.

The second, the third, and all the other molecules also have equal chances of being in the right or in the left part of the room regardless of where the others are.[8] Thus the problem of distributing molecules between the two halves of the room is equivalent to the problem of heads-and-tails distribution in a large number of tosses, and as you have seen from Figure 84, the fifty-fifty distribution is in this case by far the most probable one. We also see from that figure that with the increasing number of tosses (the number of air molecules in our case) the probability at 50 per cent becomes greater and greater, turning practically into a certainty when this number becomes very large. Since in the average-size room there are about 10^{27} molecules,[9] the probability that all of them collect simultaneously in, let us say, the right part of the room is:

$$(\tfrac{1}{2})^{10^{27}} \cong 10^{-3 \cdot 10^{26}}$$

i.e., 1 out of $10^{.3 \cdot 10^{26}}$

On the other hand, since the molecules of air moving at the speed of about 0.5 km per second require only 0.01 sec to move from one end of the room to the other, their distribution in the room will be reshuffled 100 times each second. Consequently the waiting time for the right combination is $10^{299,999,999,999,999,999,999,999,998}$ sec as compared with only 10^{17} sec representing the total age of the universe! Thus you may go on quietly reading your book without being afraid of being suffocated by chance.

To take another example, let us consider a glass of water standing on the table. We know that the molecules of water, being involved in the irregular thermal motion, are moving at high speed in all possible directions, being, however, prevented from flying apart by the cohesive forces between them.

Since the direction of motion of each separate molecule is

[8] In fact, owing to large distances between separate molecules of the gas, the space is not at all crowded and the presence of a large number of molecules in a given volume does not at all prevent the entrance of new molecules.

[9] A room 10 ft by 15 ft, with a 9 ft ceiling has a volume of 1350 cu ft, or $5 \cdot 10^7$ cu cm, thus containing $5 \cdot 10^4$ g of air. Since the average mass of air molecules is $30 \times 1.66 \times 10^{-24} \cong 5 \times 10^{-23}$ g, the total number of molecules is $5 \cdot 10^4 / 5 \cdot 10^{-23} = 10^{+27}$. ($\cong$ means: approximately equal to.)

governed entirely by the law of chance, we may consider the possibility that at a certain moment the velocities of one half of the molecules, namely those in the upper part of the glass, will all be directed upward, whereas the other half, in the lower part of the glass, will move downwards.[10] In such a case, the cohesive forces acting along the horizontal plane dividing two groups of molecules will not be able to oppose their "unified desire for parting," and we shall observe the unusual physical phenomenon of half the water from the glass being spontaneously shot up with the speed of a bullet toward the ceiling!

Another possibility is that the total energy of thermal motion of water molecules will be concentrated by chance in those located in the upper part of the glass, in which case the water near the bottom suddenly freezes, whereas its upper layers begin to boil violently. Why have you never seen such things happen? Not because they are absolutely impossible, but only because they are extremely improbable. In fact, if you try to calculate the probability that molecular velocities, originally distributed at random in all directions, will by pure chance assume the distribution described above, you arrive at a figure that is just about as small as the probability that the molecules of air will collect in one corner. In a similar way, the chance that, because of mutual collisions, some of the molecules will lose most of their kinetic energy, while the other part gets a considerable excess of it, is also negligibly small. Here again the distribution of velocities that corresponds to the usually observed case is the one that possesses the largest probability.

If now we start with a case that does not correspond to the most probable arrangement of molecular positions or velocities, by letting out some gas in one corner of the room, or by pouring some hot water on top of the cold, a sequence of physical changes will take place that will bring our system from this less probable to a most probable state. The gas will diffuse through the room until it fills it up uniformly, and the heat from the top of the glass will flow toward the bottom until all the water as-

[10] We must consider this half-and-half distribution, since the possibility that *all* molecules move in the same direction is ruled out by the mechanical law of the conservation of momentum.

sumes an equal temperature. Thus we may say that *all physical processes depending on the irregular motion of molecules go in the direction of increasing probability,* and *the state of equilibrium, when nothing more happens, corresponds to the maximum of probability.* Since, as we have seen from the example of the air in the room, the probabilities of various molecular distributions are often expressed by inconveniently small numbers (as $10^{-3 \cdot 10^{26}}$ for the air collecting in one half of the room), it is customary to refer to their logarithms instead. This quantity is known by the name of *entropy,* and plays a prominent role in all questions connected with the irregular thermal motion of matter. The foregoing statement concerning the probability changes in physical processes can be now rewritten in the form: *Any spontaneous changes in a physical system occur in the direction of increasing entropy, and the final state of equilibrium corresponds to the maximum possible value of the entropy.*

This is the famous *Law of Entropy,* also known as the Second Law of Thermodynamics (the First Law being the Law of Conservation of Energy), and as you see there is nothing in it to frighten you.

The Law of Entropy can also be called the *Law of Increasing Disorder* since, as we have seen in all the examples given above, the entropy reaches its maximum when the position and velocities of molecules are distributed completely at random so that any attempt to introduce some order in their motion would lead to the decrease of the entropy. Still another, more practical, formulation of the Law of Entropy can be obtained by reference to the problem of turning the heat into mechanical motion. Remembering that the heat is actually the disorderly mechanical motion of molecules, it is easy to understand that the complete transformation of the heat content of a given material body into mechanical energy of large-scale motion is equivalent to the task of forcing all molecules of that body to move in the same direction. However, in the example of the glass of water that might spontaneously shoot one half of its contents toward the ceiling, we have seen that such a phenomenon is sufficiently improbable to be considered as being practically impossible. Thus, *although the energy of mechanical motion can go completely over into heat* (*for example, through friction*), *the heat energy can never go completely into mechanical motion.* This rules out the possibility of the so-called "perpetual motion motor of the second kind,"[11] which would extract the heat from the material bodies at normal temperature, thus cooling them down and utilizing for doing mechanical work the energy so obtained. For example, it is impossible to build a steamship in the boiler of which steam is generated not by burning coal but by extracting the heat from the ocean water, which is first pumped into the engine room, and then thrown back overboard in the form of ice cubes after the heat is extracted from it.

But how then do the ordinary steam-engines turn the heat into motion without violating the Law of Entropy? The trick is made possible by the fact that in the steam engine *only a part of the heat liberated by burning fuel is actually turned into energy,* another larger part being thrown out into the air in the form of exhaust steam, or absorbed by the specially arranged steam coolers. In this case we have two opposite changes of entropy in our system: (1) the decrease of entropy corresponding to the transformation of a part of the heat into mechanical energy of the pistons, and (2) the increase of entropy resulting from the flow of another part of the heat from the hot-water boilers into the coolers. The Law of Entropy requires only that *the total amount* of entropy of the system increase, and this can be easily arranged by making the second factor larger than the first. The situation can probably be understood somewhat better by considering an example of a 5 lb weight placed on a shelf 6 ft above the floor. According to the Law of Conservation of Energy, it is quite impossible that this weight will spontaneously and without any external help rise toward the ceiling. On the other hand it is possible to drop one part of this weight to the floor and use the energy thus released to raise another part upward.

In a similar way we can decrease the entropy in one part of our system if there is a compensating increase of entropy in its other part. In other words *considering a disorderly motion of*

---

[11] Called so in contrast to the "perpetual motion motor of the first kind" which violates the law of conservation of energy working without any energy supply.

molecules we can bring some order in one region, if we do not mind the fact that this will make the motion in other parts still more disorderly. And in many practical cases, as in all kinds of heat engines, we do not mind it.

5. *STATISTICAL FLUCTUATION*

The discussion of the previous section must have made it clear to you that the Law of Entropy and all its consequences is based entirely on the fact that in large-scale physics we are always dealing with an immensely large number of separate molecules, so that any prediction based on probability considerations becomes almost an absolute certainty. However, this kind of prediction becomes considerably less certain when we consider very small amounts of matter.

Thus, for example, if instead of considering the air filling a large room, as in the previous example, we take a much smaller volume of gas, say a cube measuring one hundredth of a micron[12] each way, the situation will look entirely different. In fact, since the volume of our cube is 10^{-18} cu cm it will contain only $\dfrac{10^{-18} \cdot 10^{-3}}{3 \cdot 10^{-23}} = 30$ molecules, and the chance that all of them will collect in one half of the original volume is $(\frac{1}{2})^{30} = 10^{-10}$.

On the other hand, because of the much smaller size of the cube, the molecules will be reshuffled at the rate of $5 \cdot 10^9$ times per second (velocity of 0.5 km per second and the distance of only 10^{-6} cm) so that about once every second we shall find that one half of the cube is empty. It goes without saying that the cases when only a certain fraction of molecules become concentrated at one end of our small cube occur considerably more often. Thus for example the distribution in which 20 molecules are at one end and 10 molecules at the other (i.e only 10 extra molecules collected at one end) will occur with the frequency of $(\frac{1}{2})^{10} \times 5 \cdot 10^{10} = 10^{-3} \times 5 \times 10^{10} = 5 \times 10^7$, that is, 50,000,000 times per second.

Thus, on a small scale, the distribution of molecules in the air is

[12] One micron, usually denoted by Greek letter *Mu* (μ), is 0.0001 cm.

far from being uniform. If we could use sufficient magnification, we should notice the small concentration of molecules being instantaneously formed at various points of the gas, only to be dissolved again, and be replaced by other similar concentrations appearing at other points. This effect is known as *fluctuation of density* and plays an important role in many physical phenomena. Thus, for example, when the rays of the sun pass through the atmosphere these inhomogeneities cause the scattering of blue rays of the spectrum, giving to the sky its familiar color and making the sun look redder than it actually is. This effect of reddening is especially pronounced during the sunset, when the sun rays must pass through the thicker layer of air. Were these fluctuations of density not present the sky would always look completely black and the stars could be seen during the day.

Similar, though less pronounced, fluctuations of density and pressure also take place in ordinary liquids, and another way of describing the cause of Brownian motion is by saying that the tiny particles suspended in the water are pushed to and fro because of rapidly varying changes of pressure acting on their opposite sides. When the liquid is heated until it is close to its boiling point, the fluctuations of density become more pronounced and cause a slight opalescence.

We can ask ourselves now whether the Law of Entropy applies to such small objects as those to which the statistical fluctuations become of primary importance. Certainly a bacterium, which through all its life is tossed around by molecular impacts, will sneer at the statement that heat cannot go over into mechanical motion! But it would be more correct to say in this case that the Law of Entropy loses its sense, rather than to say that it is violated. In fact all that this law says is that molecular motion cannot be transformed completely into the motion of large objects containing immense numbers of separate molecules. For a bacterium, which is not *much* larger than the molecules themselves, the difference between the thermal and mechanical motion has practically disappeared, and it would consider the molecular collisions tossing it around in the same way as we would consider the kicks we get from our fellow citizens in an excited crowd.

If we were bacteria, we should be able to build a perpetual motion motor of the second kind by simply tying ourselves to a flying wheel, but then we should not have the brains to use it to our advantage. Thus there is actually no reason for being sorry that we are not bacteria!

One seeming contradiction to the law of increasing entropy is presented by living organisms. In fact, a growing plant takes in simple molecules of carbon dioxide (from the air) and water (from the ground) and builds them up into the complex organic molecules of which the plant is composed. Transformation from simple to complex molecules implies the decrease of entropy; in fact, the normal process in which entropy does increase is the burning of wood and decomposition of its molecules into carbon dioxide and water vapor. Do plants really contradict the law of increasing entropy, being aided in their growth by some mysterious *vis vitalis* (life force) advocated by old-time philosophers?

The analysis of this question indicates that no contradiction exists, since along with carbon dioxide, water, and certain salts, plants need for their growth abundant sunlight. Apart from energy, which is stored in the material of growing plants and may be liberated again when the plant burns, the rays of the sun carry with them the so-called "negative entropy" (low-level entropy), which disappears when the light is absorbed by the green leaves. Thus the photosynthesis taking place in the leaves of plants involves two related processes: a) transformation of the light energy of the sun's rays into chemical energy of complex organic molecules; b) the use of low-level entropy of the sun's rays for lowering the entropy associated with the building up of simple molecules into complex ones. In terms of "order versus disorder," one may say that, when absorbed by green leaves, the sun's radiation is robbed of the internal order with which it arrives on the earth, and this order is communicated to the molecules, permitting them to be built up into more complex, more orderly, configurations. Whereas plants build their bodies from inorganic compounds, getting their negative entropy (order) from the sun's rays, animals have to eat plants (or each other) for the supply of that negative entropy, being, so to speak, second-hand users of it.

CHAPTER IX

The Riddle of Life

1. *WE ARE MADE OF CELLS*

IN OUR discussion of the structure of matter we have purposely omitted so far any reference to a comparatively small but extremely important group of material bodies that differ from all other objects in the universe by the peculiar property of *being alive*. What constitutes the important difference between living and nonliving matter? And how far are we justified in the hope that the phenomenon of life can be understood on the basis of those fundamental physical laws that successfully explain the properties of nonliving matter?

When we speak of the phenomenon of life, we usually have in mind some fairly large and complex living organism such as a tree, a horse, or a man. But to attempt to study the fundamental properties of living matter by examining such complicated organic systems as a whole would be as futile as to attempt to study the structure of inorganic matter by regarding as a whole some complicated machine such as an automobile.

The difficulties encountered in this situation are apparent when we realize that a running automobile is formed by thousands of variously shaped parts made of different materials, in different physical states. Some of them (such as the steel chassis, the copper wires, and the glass windshield) are solid; some (such as the water in the radiator, the gasoline in the tank, and the cylinder oil) are liquid; and some (such as the mixture fed from the carburetor into the cylinders) are gaseous. The first step, then, in analyzing the complex of matter known as an automobile consists in breaking it down into separate, physically homogeneous, constituent parts. Thus we find that it is composed of various metallic substances (such as steel, copper, chromium, etc.)· various vitreous substances (such as glass and plastic materials used in construction), various homogeneous liquids (such as water and gasoline) etc., etc.

Now we can push on with our analysis and find, by using the available methods of physical investigation, that the copper parts consist of separate small crystals formed by regular layers of individual copper atoms rigidly superimposed on one another; that the water in the radiator is formed by a large number of comparatively loosely packed water molecules made of 1 oxygen and 2 hydrogen atoms each; and that the carburetor mixture streaming through the valves into the cylinders consists of a swarm of free-moving molecules of atmospheric oxygen and nitrogen molecules mixed with the molecules of gasoline vapor, which in their turn are composed of carbon and hydrogen atoms.

Similarly, in analyzing a complex living organism, such as the human body, we must first break it up into separate organs such as brain, heart, and stomach and then into various *biologically homogeneous materials* that are known under the general name "tissues."

In a sense, various types of tissues represent the material from which complex living organisms are built, in the same way as mechanical devices are constructed from various physically homogeneous substances. And the sciences of anatomy and physiology, which analyze the functioning of living organisms in terms of the properties of different *tissues* from which they are built, are in this sense analogous to the science of engineering, which bases the functioning of various machines on the known mechanical, magnetic, electrical, and other properties of physical substances used in their construction.

Thus an answer to the riddle of life cannot be found merely in seeing how the tissues are assembled to compose complex organisms but in the way in which these tissues are built from separate atoms which, in the final count, compose every living organism.

It would be a great mistake to believe that a living biologically homogeneous tissue can be compared with ordinary physically homogeneous substance. In fact, a preliminary microscopic analysis of an arbitrarily chosen tissue (whether of skin, muscle, or brain) indicates that it consists of a very large number of individual units the nature of which determines more or less the properties of the entire tissue (Figure 90). These elementary structural units of living matter are usually known as "cells"; they could also be called "biological atoms" (i.e. "indivisibles") in the sense that the biological properties of a given type of tissue will be retained only so long as it contains at least one individual cell.

A muscle tissue, for example, which is cut to the size of only half of one cell, would lose all the properties of muscular contraction, and so on, exactly in the same way as a piece of magnesium wire containing only one half of a magnesium atom would no longer be magnesium metal, but rather a small piece of coal![1]

The cells forming tissues are rather small in size (measuring on the average one hundredth of a millimeter across[2]). Any

CELLS FORMING PLANT TISSUE   A CELL FROM MUSCLE TISSUE   A CELL FROM BRAIN TISSUE

FIGURE 90

Various types of cells.

familiar plant or animal must be composed of an extremely large number of separate cells. The body of a mature human being, for example, is made of several hundred thousand billions of separate cells!

Smaller organisms are made of course of a smaller number of

[1] It will be remembered from the discussion of atomic structure that a magnesium atom (atomic number 12, atomic weight 24) consists of the nucleus formed by 12 protons and 12 neutrons surrounded by an envelope of 12 electrons. By bisecting a magnesium atom we should obtain 2 new atoms, each containing 6 nuclear protons, 6 nuclear neutrons, and 6 outer electrons—in other words, 2 atoms of carbon.

[2] Sometimes individual cells attain giant sizes, as in the familiar example of the yolk of an egg, which is known to be just one cell. In these cases, however, the vital parts of the cell responsible for its life, remain of the microscopic size, the large bulk of yellow material being simply the accumulated food to serve for the development of the chick embryo.

cells; a house fly, for example, or an ant contains no more than a few hundred million cells. There is also a large class of *monocellular* organisms, such as *amebae, fungi* (such as those causing the "ringworm" infection), and various types of *bacteria,* which are formed from one cell only, and can be seen only through a good microscope. The study of these individual living cells, which are undisturbed by "social functions" they would have to carry on in a complex organism, represents one of the most exciting chapters of biology.

In order to understand the problem of life in general, we must look for the solution in the structure and properties of the living cells.

What are the properties of living cells that make them so different from ordinary inorganic materials, or, for that matter, from the dead cells that form the wood of your writing desk or the leather in your shoes?

The fundamental distinguishing properties of the living cell consist in its abilities: (1) to assimilate materials necessary to its structure from the surrounding medium, (2) to turn these materials into the substances used for the growth of its body, and (3) to divide into two similar cells each half of its own size (and capable of growth) when its geometrical dimensions become too large. These abilities "to eat," "to grow" and "to multiply" are, of course, common to all more complex organisms made up of individual cells.

A reader with a critical mind may object by saying that these three properties can also be found in ordinary inorganic substances. If, for example, we drop a small salt crystal into a supersaturated salt solution in water,[3] the crystal will *grow* by adding to its surface successive layers of salt molecules extracted (or rather "kicked out") from the water. We can even imagine that, because of some mechanical effects, as for example the

[3] A supersaturated solution can be prepared by dissolving a large amount of salt in hot water, and cooling it to room temperature. Since solubility in water decreases with decreasing temperature, there will be more salt molecules in the water than the water can possibly keep in solution. However, the excess molecules of salt will remain in solution for a very long time unless we put in a little crystal that, so to speak, gives the initial impulse and serves as a kind of organizing agent for the exodus of salt molecules from the solution.

increasing weight of growing crystals, they will break up into two halves after reaching a certain size, and that the "baby crystals" so formed will continue the process of growing. Why shouldn't we also classify this process as a "life-phenomenon"?

In answering this, and similar questions, it must be stated first of all that, considering life simply as a more complicated case of ordinary physical and chemical phenomena, we should not expect to have a sharply defined boundary between the two cases. Similarly, using statistical laws for describing the behavior of a gas formed by an extremely large number of separate molecules (see Chapter VIII), we cannot determine the exact validity limits for such a description. In fact, we know that the atmospheric air filling the room is not going to collect itself suddenly in one corner of the room, or at least that the chances that such an unusual event will come about are negligibly small. On the other hand, we also know that if there were only two, or three, or four molecules in the entire room, they would all come to one corner rather often.

Where is the exact boundary between the number to which one statement applies and that to which the other is applicable? A thousand molecules? A million? A billion?

Similarly descending to the elementary living processes we cannot expect to find a sharp boundary between such a simple molecular phenomenon as the crystallization of salt in a water solution, and the much more complicated, though not basically different, phenomenon of the growth and division of a living cell.

In respect to this particular example we can say, however, that the growth of a crystal in a solution should not be considered as the phenomenon of life because the "food" which the crystal uses for its growth is assimilated into its body without change from its form in the solution. The molecules of salt that were previously mixed with water molecules simply collect themselves on the surface of the growing crystal. We have here an ordinary *mechanical accretion* of material instead of a typical *biochemical assimilation*. Also the multiplication of crystals, by occasional breaking up into irregular parts of no predetermined proportions, as a result of the sheer mechanical force of weight, has but little resemblance to the precise and consistent biological division of living cells into halves, which is brought about chiefly by internal forces.

We should have a much closer analogue of a biological process if, for example, the presence of a single alcohol molecule (C_2H_5OH) in a water solution of carbon dioxide gas should start a self-supporting synthetizing process that would unite one by one the H_2O-molecules of water with the CO_2-molecules of the dissolved gas, forming new molecules of alcohol.[4] Indeed, if one drop of whiskey, put into a glass of ordinary soda water should begin to turn this soda water into pure whiskey, we should be forced to consider alcohol as living matter!

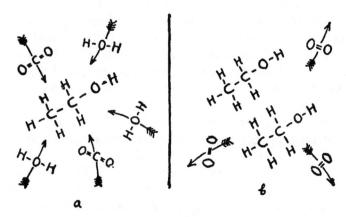

FIGURE 91

A schematic picture of the way in which an alcohol molecule *could* organize the molecules of water and carbon dioxide into another molecule of alcohol. If this process of "autosynthesis" of alcohol *were* possible we *should* have to consider alcohol as living matter.

This example is not as fantastic as it seems, since, as we shall see later, there actually exist complicated chemical substances known as *viruses*, whose rather complex molecules (formed by hundreds of thousands of atoms each) actually perform the task of organizing other molecules from the surrounding medium into

[4] For example according to the hypothetical reaction:
$$3H_2O + 2CO_2 + [C_2H_5OH] \rightarrow 2[C_2H_5OH] + 3O_2$$
where the presence of one alcohol molecule leads to the formation of another one.

structural units similar to themselves. These virus particles must be considered as ordinary chemical molecules, and as living organisms at the same time, thus *representing the "missing link" between living and nonliving matter.*

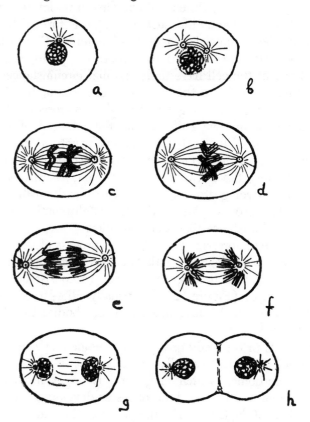

FIGURE 92

Successive stages of cell division (mitosis).

But we must return now to the problem of growth and multiplication of ordinary cells, which even though very complex are still much less so than molecules, and must be considered rather as the simplest living organisms.

If we look at a typical cell through a good microscope we see that it is made of semitransparent jellied material that has a very complicated chemical structure. It is known under the general name of *protoplasm.* It is surrounded by the cell walls which are thin and flexible in animal cells, but thick and heavy in the cells of different plants, giving to their bodies a high degree of rigidity (*cf.* Figure 90). Each cell contains in its interior a small spherical body known as the *nucleus,* which is formed by a fine network of the substance known as *chromatin* (Figure 92). It must be noticed here that various parts of the protoplasm forming the body of the cell have under normal circumstances equal optical transparencies so that the structure cannot be observed simply by looking at a living cell through a microscope. In order to see the structure we have to *dye* the material of the cell, taking advantage of the fact that different structural parts of the protoplasm absorb the dyeing materials in various degrees. The material forming the network of the nucleus is especially susceptible to the dyeing process, and appears clearly visible against the lighter background.[5] Hence the name "chromatin," which in Greek means "substance that takes on color."

When the cell is preparing for the vital process of division, the structure of the nuclear network becomes more greatly differentiated than it was before, and is seen to consist of a set of separate particles (Figure 92*b, c*), usually fiber-shaped or rodlike, which are called "chromosomes" (i.e., "bodies that take on color"). Plate V A, B.[6]

All the cells in the body of a given biological species (except the so-called reproductive cells) contain exactly the same number of chromosomes, which in general is larger in highly developed organisms than in those of lesser developed ones.

The little fruit fly, that carries the proud Latin name *Drosophila*

[5] You can use a similar method by writing something on a piece of paper with a wax candle. The writing will be invisible until you try to shade the paper with a black pencil. Since the graphite will not stick to the places covered with wax, the writing will stand out clearly on the shaded background.

[6] It must be remembered that in applying the dyeing process to a living cell we usually kill it and thus stop its further development. Thus continuous pictures of cellular divisions, such as those in Figure 92, are obtained not by observing a single cell, but by the method of dyeing (and killing) different cells in different stages of their development. In principle, however, this does not make much difference.

melanogaster and has helped the biologists to understand many things concerning the basic riddles of life, has in each of its cells *eight* chromosomes. The cells of the pea plant have *fourteen* chromosomes, and those of corn *twenty*. The biologists themselves, as well as all other people, proudly carry *forty-eight* chromosomes in each cell; which might be considered as purely arithmetical proof that man is six times better than a fly, were it not that such reasoning would prove that a crayfish, whose cells contain *two hundred* chromosomes, is more than four times better than a man!

The important thing about the number of chromosomes in the cells of various biological species is that it is *always even*; in fact in every living cell (with the exception discussed later in this chapter) we have *two almost identical sets of chromosomes* (see Plate Va): *one set from the mother and one from the father. These two sets coming from both parents carry with them the complex hereditary properties which are passed on from generation to generation of all living things.*

The initiative in cell division is taken by the chromosomes, each of which splits neatly along its entire length into two identical but somewhat thinner fibers while the cell as a whole remains intact as a single unit (Figure 92*d*).

About the time the originally tangled bundle of nuclear chromosomes begins to be organized in preparation for the division, two points known as *centrosomes*, located close to each other and near the outer boundary of the nucleus, gradually move away from each other to the opposite ends of the cell (Figure 92*a, b, c*). There also appear to be present thin threads connecting these parted centrosomes with the chromosomes inside the nucleus. When the chromosomes split into two, each half is attached to the opposite centrosome, and is firmly pulled away from the other by the contraction of the threads (Figure 92*e, f*). When this process is nearly completed (Figure 92*g*) the walls of the cell begin to cave in (Figure 92*h*) along a central line, a thin wall grows across the body of each half of the cell, the two halves let go of each other, and there are two distinct, newly produced cells.

If the two baby cells receive sufficient food from outside they will grow to their mother's size (factor of 2) and after a certain rest period undergo further division, following exactly the same pattern as that that gave them separate entities.

This description of the separate steps of cell division is the result of direct observation, and is about as far as science has been able to go in its attempts to explain the phenomenon, for very little has been observed in the direction of understanding the exact nature of the physico-chemical forces which are responsible for the process. The cell as a whole seems still to be too complicated for direct physical analysis, and before attacking this problem one must understand the nature of chromosomes— a problem somewhat simpler by comparison, and one which will be discussed in the following section.

But first it will be useful to consider how cell division is responsible for the reproductive processes in complex organisms formed by a large number of cells. Here we might be tempted to ask which came first—the chicken or the egg? But the truth is that in describing a cyclical process such as this, it doesn't matter whether we start with an "egg" that is going to develop into a chicken (or other animal), or with a chicken that is going to lay an egg.

Suppose we start with a "chicken" that has just come out of the egg. At the moment of its hatching (or birth), the cells in its body are going through a process of successive division thus effecting a rapid growth and development of the organism. Remembering that the body of a mature animal contains many thousand billion cells, all of which have been formed by successive divisions of a single fertilized egg cell, it would be natural, at first thought, to believe that in order to achieve this result there must have been a very large number of successive division-processes. One has only to remember, however, how many grains of wheat Sissah Ben induced a grateful king inadvertently to promise him by agreeing to calculate the amount on the basis of 64 simple steps in geometrical progression, or how many years would be necessary to rearrange the 64 discs of the End-of-the-World problem discussed in Chapter I, to see that comparatively few successive cell divisions would result in a very large number of cells indeed. If we designate the number of successive cell divisions necessary to the growth of a mature human being by x, and then remember that in each division the number of cells in the growing body is doubled (since each cell becomes two), we may arrive at the total number of divisions that occur in the human body between the time of the formation of the single egg cell and maturity by means of the equation: $2^x = 10^{14}$, and find that $x = 47$.

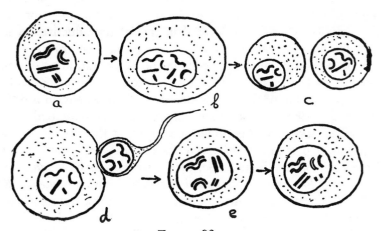

FIGURE 93

Formation of gametes (a, b, c), and fertilization of the egg cell (d, e, f). In the first process (meiosis) the paired chromosomes of the reserved reproductive cell are separated into two "half cells" without preliminary splitting. In the second process (syngamy) the male sperm cell penetrates the female egg cell and their chromosomes are paired. As a result, the fertilized cell begins to prepare for regular division, as shown earlier in Figure 92.

Thus we see that each cell in our mature bodies is a member of approximately the fiftieth generation of the original egg cell that was responsible for our existence.[7]

Although in the young animal the cells divide rather rapidly, most of the cells of a mature individual are normally in the "rest

[7] It is interesting to compare this calculation and its result with a similar calculation pertaining to the explosion of an atomic bomb (see Chapter VII). The number of successive atomic division processes needed to cause fission ("fertilization") of every uranium atom in one kilogram of material (altogether $2 \cdot 5 \cdot 10^{24}$ atoms) is calculated by the similar equation: $2^x = 2 \cdot 5 \cdot 10^{24}$, yielding $x = 61$.

state," and divide only occasionally to secure the "upkeep" of the body during life and to compensate for wear and tear.

Now we come to a very important special type of cell division which leads to the formation of the so-called "gametes" or "marrying cells" which are responsible for the reproductive phenomenon.

At the very earliest stage of any bisexual living organism, a number of its cells are set apart "in reserve" for future reproductive activity. These cells, located in special reproductive organs, undergo many fewer ordinary divisions during the growth of the organism than do any other cells in the body, and are fresh and unexhausted when they are called upon to produce new offspring. Also the division of these reproductive cells proceeds, in a different, much simpler, way than that of the ordinary body cells described above. The chromosomes forming their nuclei do not split into two as in the ordinary cells, but are simply pulled apart from each other (Figure 93*a*, *b*, *c*), so that each *daughter-cell receives only one half of the original set of chromosomes.*

The process leading to the formation of these "chromosome-deficient" cells is known as "meiosis," in contrast to the ordinary division process known as "mitosis." The cells resulting from such a division are known as "sperm cells" and "egg cells" or as the *male* and *female gametes.*

The attentive reader may wonder how the division of the original reproductive cell into two equal parts can give rise to gametes with either male or female properties, and the explanation lies in the earlier-mentioned exception to the statement that chromosomes exist only in identical pairs. There is one particular chromosome pair whose two components are identical in the female body, but different in the male. These particular chromosomes are known as sex chromosomes and are distinguished by the symbols X and Y. The cells in the body of a female have always two X-chromosomes, whereas the male has one X and one Y.[8] The substitution of a Y-chromosome for one of the X's represents the basic difference between the sexes (Figure 94).

Since all reproductive cells reserved in a female organism

[8] This statement is true for human beings and for all mammals. In birds, however, the situation is reversed; a cock has two identical sex chromosomes, whereas a hen has two different ones.

have a complete set of X-chromosomes, when one breaks in two in the process of meiosis, each half cell or gamete receives one X-chromosome. But since male reproductive cells each have one X- and one Y-chromosome, when one of them divides the result is two gametes one of which contains an X- and the other a Y-chromosome.

When, in the process of fertilization, a male gamete (sperm cell) unites with a female gamete (egg cell) there is a fifty-fifty chance that the union will result in a cell with two X-chromo-

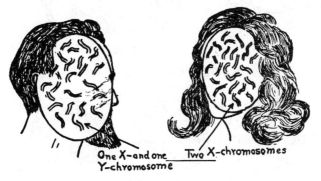

One X-and one Y-chromosome   Two X-chromosomes

FIGURE 94

The face value difference between the man and the woman. Whereas all cells of a woman's body contain 48 paired chromosomes identical within each pair, the cells of a man's body contain one asymmetric pair. Instead of two X-chromosomes as in a woman, the man has one X- and one Y-chromosome.

somes or with one X- and one Y-chromosome; in the first case the child will be a girl, in the second a boy.

We shall return in the next section to this important problem, and will proceed now with the description of the reproductive process.

When the male sperm cell unites with the female egg cell, a process known as "syngamy," there is formed a complete cell, which begins to divide into two in the process of "mitosis," illustrated in Figure 92. The two new cells so formed again divide into two each, after a short rest period; each of the four so formed repeat the process, etc. Each daughter cell receives an exact replica of all the chromosomes from the original

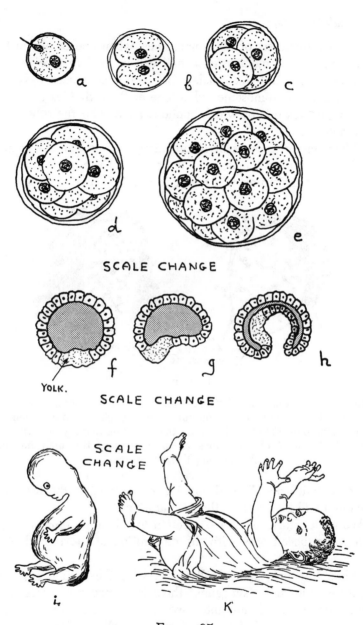

SCALE CHANGE

YOLK.

SCALE CHANGE

SCALE CHANGE

FIGURE 95
From an egg cell to the man.

fertilized egg of which half comes from the mother and half from the father. The gradual development of the fertilized egg into the mature individual is represented schematically in Figure 95. In (*a*) we see the sperm penetrating the body of a resting egg cell.

The union of two gametes stimulates a new activity in the completed cell, which now breaks up first into 2, then into 4, then into 8, then into 16, etc., etc. (Figure 95*b*, *c*, *d*, *e*). When the number of individual cells becomes rather large they tend to arrange themselves in such a way that all of them will be on the surface where they are in a better position to get food from the surrounding nutritious medium. This stage of the development in which the organism looks like a little bubble with an internal cavity is known as "bastula" (*f*). Later on, the wall of the cavity begins to bend in (*g*), and the organism enters the stage known as "gastrula" (*h*), during which it looks like a little pouch with the opening serving both for taking in fresh food and ejecting the waste from the digested materials. Simple animals such as, for example, corals never progress beyond this stage of development. In the more advanced species, however, the process of growth and advancing modification continues. Some of the cells develop into a bony skeleton, others into digestive, respiratory, and nervous systems, and going through various embryonic stages (*i*), the organism finally becomes a young animal recognizable as a member of its species (*k*).

As mentioned above, some of the developing cells of the growing organism are, even during the early stages of development, being put aside, so to speak, to be reserved for the future reproductive function. When the organism reaches maturity, these cells undergo the process of meiosis, and produce the gametes, which start the whole process over again from the beginning. And thus life marches on.

2. *HEREDITY AND GENES*

The most remarkable feature of the reproduction process lies in the fact that the new organism begotten by the union of a pair of gametes from two parents does not grow up into just any sort of living being, but develops into a rather faithful, though not necessarily exact, replica of its parents, and its parents' parents.

In fact we can be sure that a puppy born to a couple of Irish setters not only will be formed like a dog instead of like an elephant or a rabbit, but also that it will not grow as big as an elephant or remain as small as a rabbit, and that it will have four legs, one long tail, two ears and two eyes one on each side of its head. We can also be reasonably sure that its ears will be soft and hanging, that its fur will be long and golden brown in color, and that it will most probably like hunting. Besides, there will be a number of various minor points that can be traced to its father, its mother, or, maybe, to one of its earlier ancestors, and also some individual features of its own.

How were all these varied characteristics, which make up a fine Irish setter, carried within the microscopic bits of matter that made up the two gametes the union of which started the development of our puppy?

As we have seen above, every new organism receives exactly one half of its chromosomes from its father, the other half from the mother. It is clear that the major characteristics of a given species must have been contained in both the paternal and the maternal chromosomes, whereas different minor properties that can vary from individual to individual may have come separately from only one of the parents. And though there is little doubt that over a long period, and after a very large number of generations, even most of the basic properties of various animals and plants may be subject to change (organic evolution being the evidence for that), it is only the comparatively small changes of minor characteristics that can be noticed during the limited period of observation that is represented by human knowledge.

The study of such characteristics and their transfer from parents to children is the main subject of the new science of *genetics* which, being still practically in its infancy, is able nevertheless to tell us very exciting stories about the most intimate secrets of life. We have learned, for instance, that, in contrast to most biological phenomena, the laws of heredity possess almost mathematical simplicity, indicating that we are dealing here with one of the fundamental phenomena of life.

Take for example such a well-known defect of human eyesight as *color blindness,* the most common form of which is marked by an inability to distinguish between red and green. In order to explain color blindness we must first understand why we see colors at all, through study of the complicated structure and properties of the retina, the problems concerning photochemical reactions caused by light of different wavelengths, etc., etc.

But if we ask ourselves about the *inheritance of color blindness,* a question that would seem at first to be still more complicated than the explanation of this phenomenon itself, the answer is unexpectedly simple and easy. It is known from observed facts: (1) that men are much more often color-blind than are women, (2) that the children of a color-blind man and a "normal" woman are never color-blind, but (3) that among the children of a color-blind woman and a "normal" man, the sons are color-blind whereas the daughters are not. Knowing these facts, which indicate clearly that the inheritance of color blindness is somehow associated with sex, we have only to assume that the characteristics of color blindness result from a defect in one of the chromosomes and is transferred with this chromosome from generation to generation in order to combine knowledge and logical assumption in the further assumption that *color blindness results from a defect in the sex chromosome that we have previously denoted by X.*

With this assumption the empirical rules concerning color blindness become clear as crystal. Remember that the female cells possess two X-chromosomes whereas male cells possess only one (the other being the Y-chromosome). If the single X-chromosome in man is defective in this particular way he is color-blind. In a woman, both X-chromosomes must be affected since one chromosome only is enough to secure the perception of color. If the chances that an X-chromosome has this color defect are, say, one in a thousand, there will be one color-blind individual among a thousand men. The *a priori* chances that both the X-chromosomes in a woman have the color defect are calculated according to the theorem of the multiplication of probabilities, (see Chapter VIII) by the product: $\dfrac{1}{1000} \times \dfrac{1}{1000} = \dfrac{1}{1,000,000}$, so that only 1 woman out of 1,000,000 may be expected to be color-blind.

Let us consider now the case of a color-blind husband and a "normal" wife (Figure 96a). Their sons will receive no X-chromosomes from their father, and one "good" X-chromosome from their mother, thus having no reason for color blindness.

Their daughters, on the other hand, will have a "good" X-chromosome from their mother and a "bad one" from the father.

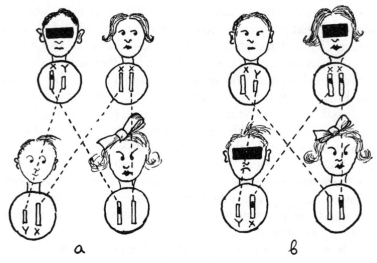

FIGURE 96

Heredity of color blindness.

They will not be color-blind, though their children (sons) may be.

In the opposite case of a color-blind wife and a "normal" husband (Figure 96b), the sons will be definitely color-blind since their single X-chromosome comes from the mother. The daughters, who will have one "good" X-chromosome from the father and one "bad" one from the mother, will not be color-blind, but as in the previous case, their sons will be color-blind. Simple as simple can be!

Such hereditary properties as color blindness, which require *both chromosomes* of the pair to be affected in order to produce a noticeable effect, are known as "recessive." They can be carried over from the grandparents to the grandchildren in a hidden form, and are responsible for such sad facts as that an occasional puppy born from two good-looking German shepherd dogs may look like anything but a German shepherd.

The opposite is true of the so-called "dominant" characteristics, which become noticeable when only one chromosome of the pair

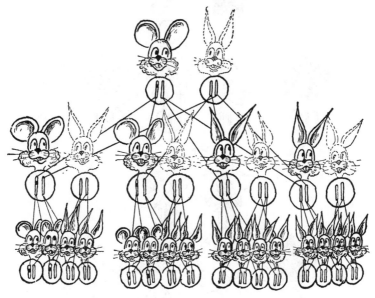

FIGURE 97

is affected. In order to get away from the factual material of genetics, we shall illustrate this case by an imaginary example of a wabbit (pardon me, a rabbit) born with ears resembling those of Mickey Mouse. If we assume that "Mickey ears" are a *dominant* characteristic in heredity, that is, that a change in one single chromosome suffices to make the ears grow in this shameful (for a rabbit) way, we can forecast the kinds of ears succeeding generations of rabbit offspring will have by looking at Figure 97, assuming that the rabbits born of the original and succeeding unions will mate with normal rabbits. The deviation from normal in the chromosome responsible for the Mickey ears is marked in our diagram by a black spot.

In addition to the *dominant* and *recessive* indented characteristics there are also those that may be called "indifferent." Suppose we have in our garden red and white four o'clocks.

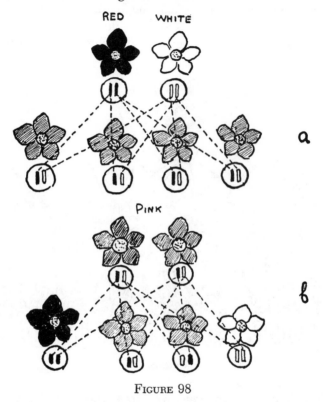

FIGURE 98

When the pollen (sperm cells in plants) from red flowering plants is carried over by the wind or insects to the pistils of another red plant, they unite with the *ovules* (egg cells in plants) located at the base of the pistil and develop into the seeds that will produce again red flowers. So if pollen from white flowers fertilizes other white flowers, the next generation of flowers will all be white. If, however, the pollen from white flowers falls on red ones, or vice versa, the plant grown from the seeds thus produced will have pink flowers. It is easy to see, however, that pink flowers do not represent a biologically stable variety. If we breed them within the group we find that the next generation will be 50 per cent pink, 25 per cent red, and 25 per cent white.

The explanation can be readily found if we assume that the property of being red or white is carried by one of the chromosomes in the plant's cell, and that in order to have a pure color, *both* chromosomes of the pair must be identical in this respect. If one chromosome is "red" whereas another is "white," the battle of colors results in pink flowers. Looking at Figure 98, which represents schematically the distribution of "color chromosomes" among the offspring, we may see the numerical relation mentioned above. It would also be easy to show, by drawing another diagram similar to that of Figure 98, that by breeding white and pink four o'clocks we should get in the first generation 50 per cent pink and 50 per cent white but no red flowers. Similarly red and pink flowers will give 50 per cent red, 50 per cent pink, but no white flowers. Such are the laws of heredity that were first discovered almost a century ago by the unpretentious Moravian monk Gregor Mendel while growing garden peas in the monastery near Bruns.

Thus far we have been associating various properties inherited by the young organism with different chromosomes that it receives from its parents. But, since there is an almost uncountable number of different properties, as compared with a comparatively small number of chromosomes (8 in each cell of a fly, 48 in each of a man), we are forced to acknowledge that each chromosome carries in it a long list of individual characteristics, which can be imagined as distributed along its thin fiberlike body. In fact, looking at Plate VA, which represents the chromosomes of the salivary glands of a fruit fly (*Drosophila melanogaster*[9]), it is difficult to escape the impression that the numerous darkish strata cutting across the long body of the chromosome represent the sites of different properties carried by it. Some of these cross bands may regulate the color of the fly, some the shape of its

[9] In this particular case, in contrast to the plurality of other cases, chromosomes are exceptionally large, and their structure can be easily studied by the methods of microphotography.

wings, and still others may be responsible for the fact that it has *six* legs, that it is about a *quarter of an inch* long, and that it looks in general like a fruit fly and not a centipede or a chicken.

And, in fact, the science of genetics tells us that this impression is quite correct. It is possible not only to show that these tiny structural units of a chromosome, known as "genes," carry in themselves various individual hereditary properties, but one can

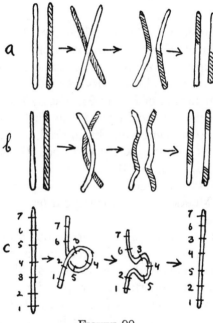

FIGURE 99

also tell in many cases which particular gene carries one or another particular property.

Of course, even with the largest possible magnification, all genes look almost alike, their functional difference being hidden somewhere deep in their molecular structure.

Thus their individual "purpose in life" can be found only by careful studies of the way in which different hereditary properties are carried from generation to generation in a given species of plants or animals.

We have seen that any new living organism gets half of its

chromosomes from the father and half from the mother. Since the paternal and maternal sets of chromosomes represent the 50-50 mixture of those coming from corresponding grandparents, we should expect that the child gets its heritage from only one of the grandparents on each side. This, however, is known not to be necessarily true, and there are cases in which *all four grandparents bequeath characteristics to their grandchild.*

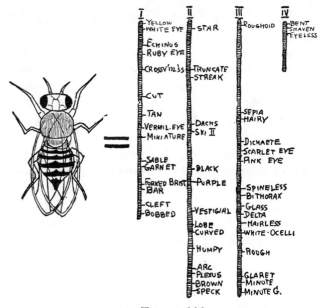

FIGURE 100

Does this mean that the scheme of chromosome-transfer described above is wrong? No, it is not wrong, but only somewhat simplified. The factor that is still to be taken into account is that, in preparation for the process of meiosis, by which the reserved reproductive cell breaks up into two gametes, the paired chromosomes often become twisted around each other, and can exchange their parts. Such exchange processes, shown schematically in Figure 99*a*, *b*, lead to the mixture of gene sequences obtained from the parents, and are responsible for hereditary mix-ups. There are also cases (Figure 99*c*) in which a single chromosome can be folded into a loop, and may then break up in a different way, mixing up the order of genes in it (Figure 99c; Plate VB).

It is clear that such reshuffling of genes between two chromosomes of a pair, or within a single chromosome, will more probably affect the relative positions of those genes that were originally far apart than those that were close neighbors. Exactly in the same way, cutting a pack of cards will change the relative positions of the cards below and above the cut (and will bring together the card that was at the top of the pack and that card that was at the bottom) but will separate only one pair of immediate neighbors.

Thus, observing that two definite hereditary properties almost always travel together in the crossing-over of chromosomes, we may conclude that the corresponding genes were close neighbors. On the contrary, the properties often separated in the crossingover process must be located in the distant parts of the chromosome.

Working along these lines, the American geneticist T. H. Morgan and his school were able to establish a definite order of genes in the chromosomes of the fruit fly, which was used in their investigations. Figure 100 is a chart showing how different characteristics of the fruit fly are distributed in the genes of the four chromosomes that make up the fruit fly, as discovered by such research.

A chart such as that shown in Figure 100, which was made for the fly, could of course also be made for more complicated animals including man although this would require much more careful and detailed studies.

3. GENES AS "LIVING MOLECULES"

Analyzing step by step the immensely complicated structure of living organisms, we have now come to what seems to be the *fundamental units of life*. In fact, we have seen that the entire course of development and practically all the properties of the grown-up organism are regulated by a set of genes hidden in the deep interior of its cells; one may say that every animal or plant "grows around" its genes. If a highly simplified physical analogy is to be permitted here, we may compare the relation between the genes and the living organism with the relation between the atomic nuclei and the large lumps of inorganic matter. Here too, practically all physical and chemical properties of a given substance can be reduced to the basic properties of the atomic nuclei that are characterized simply by one number designating their electric charge. Thus, for example, the nuclei carrying a charge of 6 elementary electric units will surround themselves by the atomic envelopes of 6 electrons each, which will give these atoms a tendency to arrange themselves in a regular hexagonal pattern, and to form the crystals of exceptional hardness and very high refractive index that we call diamonds. Similarly a set of nuclei with electric charges 29, 16, and 8 will give rise to the atoms that stick together to form soft blue crystals of the substance known as copper sulfate. Of course, even the simplest living organism is much more complicated than any crystal, but in both cases we have the typical phenomenon of macroscopic organization being determined to the last detail by microscopic centers of organizing activity.

How large are these organizing centers that determine all the properties of living organisms from the fragrance of a rose to the shape of the elephant's trunk? This question can easily be answered by dividing the volume of a normal chromosome by the number of genes contained in it. According to microscopic observations an average chromosome is about a thousandth of a millimeter thick, which means that its volume is about 10^{-14} cu cm. Yet breeding experiments suggest that one chromosome must be responsible for as many as several thousand different hereditary properties, a figure that can also be directly obtained by counting the number of dark bands (presumably separate genes) crossing the long bodies of the enormously overgrown chromosome of the fruit fly *Drosophila melanogaster*[10] (Plate V). Dividing the total volume of the chromosome by the number of separate genes we find that the volume of one gene is not larger than 10^{-17} cu cm. Since the volume of an average atom is about 10^{-23} cu cm $[\cong (2 \cdot 10^{-8})^3]$, we conclude that *each separate gene must be built from about one million atoms*.

We can also estimate the total weight of genes, say, in the

---

[10] The normal size chromosomes are so small that microscopic investigation fails to resolve them into separate genes.

body of a man. As we have seen above, a grown-up person is made of about 10^{14} cells, each cell containing 48 chromosomes. Thus the total volume of all chromosomes in the human body is about $10^{14} \times 48 \times 10^{-14} \cong 50$ cm³ and (since the density of living substance is comparable with the density of water) it must weigh less than two ounces. It is this almost negligibly small amount of "organizing substance" that builds around itself the complicated "envelope" of the animal's or plant's body thousands of times its own weight, and that rules "from inside" every single step of its growth, every single feature of its structure, and even a very large part of its behavior.

But what is the gene itself? Must it be considered as a complicated "animal," which could be subdivided into even smaller biological units? The answer to this question is definitely no. The gene is the smallest unit of living matter. Further, while it is certain that genes possess all of those characteristics that distinguish matter possessing life from matter that does not, there is also hardly any doubt that they are linked on the other side with the complex molecules (like those of proteins), which are subject to all the familiar laws of ordinary chemistry.

In other words, it seems that *in the gene we have the missing link between organic and inorganic matter, the "living molecule" that was contemplated in the beginning of this chapter.*

Indeed, considering on the one hand the remarkable permanence of genes, which carry almost without any deviation the properties of a given species through thousands of generations, and on the other hand the comparatively small number of individual atoms that form one gene, one cannot consider it otherwise than as a well-planned structure in which each atom or atomic group sits in its predetermined place. The differences between the properties of various genes, which are reflected in the external variations among resulting organisms, the characteristics of which they determine, can then be understood as due to variations in the distribution of the atoms within the structure of the genes.

To give a simple example, let us take a molecule of TNT (Trinitrotoluene), an explosive material, which played a prominent role in the last two wars. A TNT molecule is built from 7 carbon atoms, 5 hydrogen atoms, 3 nitrogen atoms, and 6 oxygen atoms, arranged according to one of the schemes:

```
        H                        H                        H
        |                        |                        |
     H–C–H                    H–C–H                    H–C–H
        |                        |                        |
        C                        C                        C
  O   /    \\   O        O   /    \\   O               /    \\
   >N–C      C–N<         >N–C      C–N<       H–C      C–N<  O
  O    |     |   O       O    ||     |   O              |      O
     H–C     C–H           H–C      C–N<       H–C      C–N<  O
        \    /                \    /   O           ||   /      O
         C                     C                    C
         |                     |                    |
         N                     H                    N
       // \\                                      // \\
      O     O                                    O     O

        α                        β                        γ
```

The differences between the three arrangements lies in the way

in which $N{\Large{<}}^O_O$ groups are attached to the C-ring, and the result-

ing materials are usually designated as αTNT, βTNT, and γTNT. All three substances can be synthesized in the chemical laboratory. All three are explosive in nature, but show small variations in density, solubility, melting point, explosive power, etc. Using the standard methods of chemistry, one can easily transplant the

$N{\Large{<}}^O_O$ groups from one set of points of attachment within the

molecule to another, thus changing one brand of TNT into another. Examples of such kind are very common in chemistry, and the larger the molecule in question the larger the number of the varieties (*isomeric forms*) which can be thus produced.

If we consider the gene as one giant molecule built from a million atoms, the number of possibilities for arranging various atomic groups in different places within the molecule becomes immensely large.

We may think of the gene as a long chain composed of periodically repeating atomic groups with various other groups attached to it, as pendants are attached on a charm bracelet; indeed, recent advances in biochemistry permit us to draw an exact diagram of that hereditary charm bracelet. It is formed of the atoms of carbon, nitrogen, phosphorus, oxygen, and hydrogen, and is known as ribonucleic acid. In Fig. 101 we give a somewhat surrealistic picture (with nitrogen and hydrogen atoms omitted) of the part of the hereditary bracelet that determined the color of a newborn child's eyes. The four pendants tell that the baby's eyes are gray.

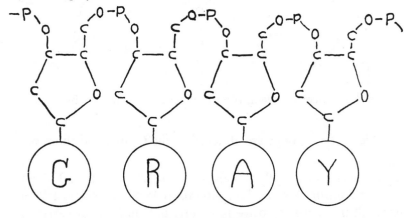

FIGURE 101

A part of hereditary "charm bracelet" (the molecule of ribonucleic acid) determining the color of eyes. (Highly schematic!)

In transposing the different pendants from one hook to another we can get an almost infinite variation of different distributions.

Thus, for example, having a charm bracelet with 10 different pendants, we can distribute them in $1 \times 2 \times 3 \times 4 \times 5 \times 6 \times 7 \times 8 \times 9 \times 10 = 3,628,800$ different ways.

If some of the pendants are identical, the number of possible arrangements will be smaller. Thus, having only 5 kinds of pendants (2 of each), we will have only 113,400 different possibilities. The number of possibilities increases, however, very rapidly with the total number of pendants, and if we have, for example,

25 pendants, 5 of each different kind, the number of possible distributions is approximately 62,330,000,000,000,000!

Thus we see that the number of different combinations that can be obtained by redistribution of different "pendants" among various "suspension-places" in long organic molecules is certainly large enough to account not only for all the varieties of known living forms but also for the most fantastic nonexistent forms of animals and plants which can be created by our imagination.

A very important point concerning the distribution of property-characterizing pendants along the fiberlike gene molecules is that this distribution is subject to spontaneous changes resulting in corresponding macroscopic changes in the entire organism. The most common cause for such changes lies in the ordinary thermal motion, which makes the entire body of the molecule bend and twist like the branches of a tree in a strong wind. At sufficiently high temperatures this vibrational motion of molecular bodies becomes sufficiently strong to break them up into separate pieces —a process known as thermal dissociation (see Chapter VIII). But, even at lower temperatures, when the molecule as a whole retains its integrity, thermal vibrations may result in some internal changes of molecular structure. We can imagine, for example, that the molecule is twisted in such a way that one of the pendants attached at one point is brought close to some other point of its body. In such a case it may easily happen that the pendant will get disconnected from its previous location, and become attached at a new spot.

Such phenomena, called *isomeric transformations*,[11] are well known in ordinary chemistry in cases of comparatively simple molecular structures, and, along with all other chemical reactions, follow the fundamental law of chemical kinetics according to which *the rate of reaction increases approximately by a factor of 2 each time the temperature rises by 10° C.*

In the case of *genes molecules*, the structure of which is so complex that it will probably defy the best efforts of organic chemists for a long time to come, there is at present no way to

---

[11] The term "isomeric," as has already been explained, refers to the molecules that are built from the same atoms, which are, however, arranged in different manner.

certify isomeric changes by the direct methods of chemical analysis. However, we have in this case something that, from a certain point of view, can be considered much better than laborious chemical analysis. If such an isomeric change takes place in one of the genes inside a male or female gamete the union of which is going to give rise to a new living organism, it will be faithfully repeated in the successive processes of gene splitting and cell division, and will affect some easily observable macroscopic features of the animal or plant thus produced.

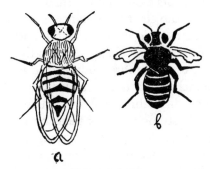

FIGURE 102

Spontaneous mutation of a fruit fly.
(*a*) Normal type: gray body, long wings.
(*b*) Mutated type: black body, short (vestigial) wings.

And indeed, one of the most important results of genetical studies lies in the fact (discovered in 1902 by the Dutch biologist de Vries) that *spontaneous hereditary changes in living organisms always take place in the form of discontinuous jumps known as mutations.*

To give an example, let us consider the breeding experiments of fruit flies (*Drosophila melanogaster*) that have already been mentioned. The wild variety of fruit flies have gray bodies and long wings; and whenever you catch one in the garden you may be almost completely sure that it will fill these specifications. However, breeding generation upon generation of these flies under laboratory conditions, one obtains once in a while a peculiar "freak type" of fly with abnormally short wings and an almost black body (Figure 102).

The important point is that you will probably not find along with the short-winged black fly other flies with various shades of gray and with wings of varying length, standing in successive stages of variation between the extreme exception (nearly black body and very short wings) and the "normal" ancestors, in a progression of increasingly modified generations. As a rule, all the members of a new generation (and there may be hundreds of them!) are about equally gray with equally long wings, and only one (or a few) are *entirely* different. *Either there is no substantial change, or there is quite a large change (mutation).* A similar situation has been observed in hundreds of other cases. Thus, for example, color blindness does not necessarily come through heredity, and there must be cases where a baby is born color-blind without any "guilt" on the part of its ancestors. In the case of color blindness in men, just as in the case of short wings in the fruit fly, we have the same principle of "all or nothing"; it is not a question of a person's being better or worse in distinguishing the two colors—either he does or he doesn't.

As everybody who ever heard the name of Charles Darwin knows, these changes in the properties of new generations combined with *the struggle for existence* and *the survival of the fittest* lead to the steady process of *evolution of species*,[12] and are responsible for the fact that a simple mollusk, who was the king of nature a couple of billion years ago, has developed into a highly intelligent living being like yourself, who is able to read and to understand even such a highly sophisticated book as this one.

The jumplike variations in hereditary properties is perfectly understandable from the point of view of isomeric changes in gene molecules as discussed above. In fact, if the property-determining pendant in a gene molecule changes its place, it cannot do so halfway; either it remains in the old place, or it becomes attached to a new place, thus causing a discontinuous change in the properties of the organism.

The point of view according to which "mutations" are due to

---

[12] The only difference that the discovery of mutations has introduced into Darwin's classical theory is that evolution is due to discontinuous jumplike changes and not to continuous small changes as Darwin had in mind.

isomeric changes in the gene molecules is given strong support by the way in which the rate of mutations depends on the temperature of the enclosure in which the animals or plants are bred. In fact, the experimental work of Timoféëff and Zimmer on the effect of temperature on the rate of mutations indicates that (apart from some additional complications caused by the surrounding medium and other factors) it follows the same fundamental physicochemical law as any other ordinary molecular reaction. This important discovery caused Max Delbrück (formerly a theoretical physicist, now an experimental geneticist) to develop his epoch-making views concerning the equivalence between the biological phenomenon of mutations and the purely physicochemical process of isomeric changes in a molecule.

We could go on indefinitely discussing the physical foundation of the gene theory, in particular the important evidence supplied by the study of mutations produced by X rays and other radiation, but what has already been said appears to be sufficient to convince the reader of the fact that *science is at present crossing the threshold of the purely physical explanation of the "mysterious" phenomenon of life.*

We cannot finish this chapter without reference to the biological units known as *viruses,* which seem to represent *free genes* without a cell around them. Until a comparatively recent time biologists believed that the simplest forms of life were represented by various types of *bacteria,* the unicellular microorganisms that grow and multiply in the living tissues of animals and plants, sometimes causing various kinds of diseases. Microscopic studies have revealed, for example, that typhoid fever is due to a special type of bacteria having strongly elongated bodies, about 3 microns (μ)[13] long, and $\frac{1}{2}$ μ across, whereas the bacteria of scarlet fever are spherically shaped cells about 2 microns in diameter. There were, however, a number of diseases, such as, for example, influenza in man or the so called mosaic-disease in the tobacco plant, where ordinary microscopic observations failed to discover any normal-sized bacteria. Since, however, these particular "bacterialess" diseases were known to be carried from the body of sick individuals into the body of healthy ones in the same "in-

---

[13] A micron is one thousandth of a millimeter. or 0.0001 cm.

fectious" way as all other ordinary diseases, and since the "infection" thus received spread rapidly over the entire body of the infected individual, it was necessary to assume that they were associated with some kind of hypothetical biological carriers, which received the name *virus*.

But it was only comparatively recently that the development of *ultramicroscopic technique* (using ultraviolet light) and especially the invention of the *electron microscope* (in which the use

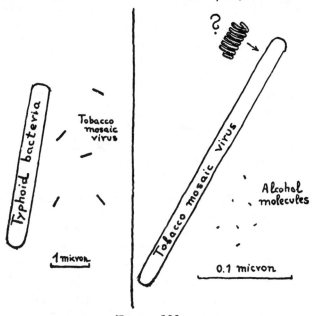

FIGURE 103

Comparison between bacteria, viruses, and molecules.

of electron beams instead of ordinary light rays permits much larger magnifications) permitted microbiologists to see for the first time the formerly hidden structure of viruses.

It was found that various viruses represent collections of a large number of individual particles, all of them of exactly the same size and much smaller than ordinary bacteria (Figure 103). Thus the particles of influenza virus are little spheres 0.1 μ in diameter, whereas the slender stick-shaped particles of tobacco-mosaic virus are 0.280 μ long and 0.015 μ across.

In Plate VI we have a very impressive electron-microscopic photograph of tobacco-mosaic virus particles, the smallest known living units in existence. Remembering that the diameter of one atom is about 0.0003 μ, we conclude that the particle of tobacco-mosaic virus measures only about fifty atoms across, and about a thousand atoms along the axis. Altogether not more than a couple of million individual atoms![14]

This figure immediately brings to our mind the similar figure obtained for the number of atoms in a single gene, and brings up the possibility that the virus particles may be considered as "free genes" that did not bother to unite in the long colonies that we call chromosomes, and to surround themselves by a comparatively ponderous mass of cellular protoplasm.

And indeed the reproductive process of the virus particles seems to go along exactly the same lines as the doubling of chromosomes in the process of cell division: their entire body splits along its axis giving rise to two new full-size virus particles. Apparently we observe here the basic reproductive process (illustrated in Figure 91 for a fictitious case of alcohol reproduction) in which various atomic groups located along the length of the complicated molecule attract from the surrounding medium similar atomic groups, arranging them in exactly the same pattern as in the original molecule. When the arrangement is completed the new molecule, already mature, splits away from the original one. In fact, it seems that in the case of these primitive living organisms the usual process of "growth" does not take place, and that the new organisms are simply built "by parts" alongside the old ones. The situation may be illustrated by imagining a human child who develops on the outside of, and attached to, the body of the mother, detaching itself and walking away when it is a completely formed man or woman. (The author will not draw a picture of such a situation, in spite of a strong temptation to do

[14] The number of atoms forming a virus particle may be actually considerably less than this, since it is quite possible that they are "empty inside" being formed by the coiled molecular chains of the type shown in Figure 101. If we assume that the tobacco-mosaic virus actually has such a structure (shown schematically in Figure 103), so that various atomic groups are located only on the *surface* of the cylinder, the total number of atoms per particle will be reduced to only a few hundred thousand. The same argument also pertains, of course, to the number of atoms in a single gene.

so.) It goes without saying that, in order to make such a multiplication process possible, the development must proceed in a special, partially organized medium; and in fact, in contrast to the bacteria that have protoplasm of their own, virus particles can multiply only within the living protoplasm of other organisms being in general very choosy about their "food."

Another common characteristic of viruses is that *they are subject to mutations* and that the mutated individuals pass the newly acquired characteristics to their offspring in accordance with all the familiar laws of genetics. In fact biologists have been able to distinguish several hereditary strains of the same virus, and to follow their "race development." When the new epidemics of influenza sweep through communities, one can be fairly sure that they are caused by some new mutated type of influenza virus having some new vicious properties against which the human organism has not yet had the chance to develop proper immunity.

In the course of the previous pages we have developed a number of strong arguments showing that *virus particles must be considered as living individuals.* We can now assert with no less vigor that *these particles must also be considered as regular chemical molecules* subject to all the laws and regulations of physics and chemistry. In fact, purely chemical studies of virus material establish the fact that a given virus may be considered as a well-defined chemical compound, and may be treated in the very same way as various complex organic (but not living) compounds, and that they are subject to various types of substitutional reactions. It seems in fact to be only a matter of time before the biological chemist will be able to write for each virus a structural chemical formula just as easily as he now writes the formula for alcohol, glycerine, or sugar. Still more striking is the fact that virus particles of a given type are apparently of *exactly the same size* up to their last atom.

In fact it was shown that virus particles deprived of the feeding medium in which they live arrange themselves into regular patterns of ordinary crystals. Thus, for example, the so-called *"tomato bushy stunt"* virus crystallizes in the form of large beautiful rhombic dodecahedrons! You can keep this crystal in a mineralogical cabinet along with feldspar and rock salt; but put it back into the tomato plant and it will turn into a swarm of living individuals.

The first important step in synthesizing living organisms from inorganic material was recently made by Heinz Fraenkel-Conrat and Robley Williams in the Virus Institute of the University of California. Working with the tobacco-mosaic virus, they managed to separate these virus particles into two parts, each of which represents a non-living, though rather complex, organic molecule. It was long known that this virus, having the shape of long sticks (plate VI), is formed by a bunch of long straight molecules of organizing material (known as *ribonucleic acid*) with long protein molecules wound around it like a coil of electric wire around the iron core in an electromagnet. By using various chemical reagents, Fraenkel-Conrat and Williams succeeded in breaking up these virus particles, separating the ribonucleic acid from the protein molecules without damaging them. Thus, they obtained in one test tube a water solution of ribonucleic acid, and in another a solution of protein molecules. Electron-microscope photographs have shown that the test tubes contained nothing but the molecules of these two substances, completely devoid of any trace of life.

But when the two solutions were put together, the molecules of ribonucleic acid began to combine into groups of 24 molecules in each bunch, while the protein molecules started to wind around them, forming exact replicas of the virus particles with which the experiment was started. When applied to the leaves of the tobacco plant, these taken-apart-and-put-together-again virus particles caused the mosaic disease in the plant, as if they had never been taken apart. Of course, in this case the two chemical components in the test tubes were obtained by breaking up a living virus. The point is, however, that biochemists are now in possession of methods for synthesizing from ordinary chemical elements both ribonucleic acid and protein molecules. Although at the present time (1960) only comparatively short molecules of both substances can be synthesized, there is no reason to doubt that in the course of time molecules as long as those in viruses will be made from simple elements. And putting them together would produce a man-made virus particle.

PART IV

Macrocosmos

CHAPTER X

Expanding Horizons

1. *THE EARTH AND ITS NEIGHBORHOOD*

NOW, returning from our excursion into the reign of molecules, atoms and nuclei, back to objects of more accustomed size, we are ready again to start a new journey but this time in the opposite direction, that is, toward the sun, the stars, the distant stellar clouds and the outflung boundaries of our

FIGURE 104
The world of the ancients

universe. Here, as in the case of the microcosmos, the development of science leads us farther and farther from the familiar ground of everyday objects and opens up gradually ever-broadening horizons.

In the early stages of human civilization, the thing that we call the universe was considered almost ridiculously small. The earth was believed to be a large flat disc floating on the surface of the world ocean which surrounded it. Below was only water as deep as one could imagine, above was the sky, the abode of the gods. The disc was large enough to hold all lands known to the geography of that time, which included the shores of the Mediterranean Sea, with the adjacent parts of Europe, Africa, and a bit of Asia. The northern rim of the Earth disc was limited by a range of high mountains, behind which the Sun hid during the night time when it was resting on the surface of the World Ocean. The picture on page 269 (Figure 104) gives a fairly accurate idea of how the world looked to the people of ancient history. But in the third century before Christ there lived a man who disagreed with this simple and generally accepted picture of the world. He was the famous Greek philosopher (so they called scientists at that time) named Aristotle.

In his book *About Heaven* Aristotle expressed the theory that our Earth is actually a sphere, covered partly by land, partly by water, and surrounded by the air. He supported his point of view by many arguments which are familiar and seem trivial to us now. He indicated that the way the ships disappear behind the horizon when the hulk vanishes first and the masts seem to stick out of the water, proves that the surface of the ocean is curved, not flat. He argued that the eclipses of the moon must be due to the shadow of the Earth passing over the face of our satellite, and since this shadow is round, the Earth itself must be round too. But only very few people at that time would believe him. People could not understand how, if what he said was true, those who lived on the opposite side of the globe (the so-called antipodes; Australians to you) could walk upside down without falling off the Earth, or why the water in these parts of the world did not flow toward what they would call the blue sky (Figure 105).

The people at that time, you see, did not realize that the things fall down because they are attracted by the body of the Earth. For them "above" and "below" were absolute directions in space, which should be the same everywhere. The idea that "up" can become "down" and "down" become "up" if you travel halfway around the Earth must have seemed to them just as

crazy as many statements of Einstein's theory of relativity seem
to many people today. The fall of heavy bodies was explained
not by the pull of the Earth, as we explain it now, but by the

FIGURE 105

An argument against the spherical shape of the Earth.

"natural tendency" of all things to move downward; and so down
you go toward the blue sky if you venture to put your foot on
the under part of the Earth globe! So strong was the objection
and so hard the adjustment to the new ideas that in many a book
published as late as the fifteenth century, almost two thousand

years after Aristotle, one could find pictures showing inhabitants of the antipodes standing head down on the "underneath" of the Earth, and ridiculing the idea of its spherical shape. Probably the great Columbus himself, setting off for his journey to discover "the-other-way-round road" to India, was not completely sure of the soundness of his plan, and as a matter of fact he did not fulfill it because the American Continent got in the way.

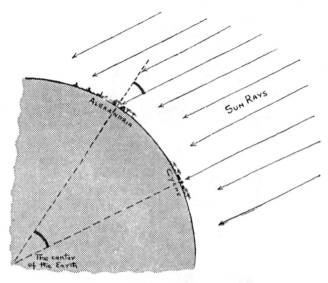

FIGURE 106

And only after the famous around-the-world sailing of Fernando de Magalhães (better known as Magellan) did the last doubt about the spherical shape of the Earth finally disappear.

When it was first realized that the Earth has the shape of a giant sphere, it was natural to ask how large this sphere was in comparison with the parts of the world known at that time. But how would you measure the size of the Earth without undertaking a round-the-world trip, which was of course out of the question for the philosophers of ancient Greece?

Well, there is a way, and it was first seen by the famous scientist of that time named Eratosthenes, who lived in the Greek colony of Alexandria in Egypt during the third century B. C. He had heard from the inhabitants of Syene, a city on the Upper Nile some 5000 Egyptian stadia south from Alexandria,[1] that during the summer solstice the noon sun in that city stood directly overhead, so that vertical objects threw no shadow. On the other hand Eratosthenes knew that no such thing ever happened in Alexandria, and that on the same day the sun passes 7 degrees, or one fiftieth of the full cycle, away from the zenith (the point directly overhead.) Assuming that the Earth is round, Erathosthenes gave a very simple explanation of that fact, an explanation that you can easily understand by looking at Figure 106. Indeed, since the surface of the earth curves between the two cities, the sun rays falling vertically in Syene are bound to strike the earth at a certain angle in the more northerly located Alexandria. You can also see from that figure that if two straight lines were drawn from the center of the earth, one to pass through Alexandria and one through Syene, the angle that they would make at their convergence would be identical with that made by the convergence of the line passing from the center of the earth to Alexandria (i.e. zenith direction in Alexandria) and the sun's rays at the time that the sun is directly over Syene.

Since that angle is one fiftieth of the full circle, the total circumference of the earth should be fifty times the distance between the two cities, or 250,000 stadia. One Egyptian stadium is about $\frac{1}{10}$ mile so that Eratosthenes' result is equivalent to 25,000 miles, or 40,000 km; very close indeed to the best modern estimates.

However the main point of the first measurement of the Earth was not in the exactness of the number obtained, but in the realization of the fact that the Earth was *so* large. Why, its total surface must be several hundred times larger than the area of all known lands! Could it be true, and if true, what was beyond the known borders?

Speaking about astronomical distances, we must get acquainted first with what is known as *parallactic displacement* or simply as *parallax*. The word may sound a little frightening, but as a matter of fact the parallax is a very simple as well as useful thing.

We may start our acquaintance with parallax by trying to put

[1] Near present location of the Assuan Dam.

a thread into a needle's eye. Try to do it with one eye closed, and you will find very quickly that it does not work; you will be bringing the end of the thread either too far behind the needle or stopping short in front of it. With only one eye you are unable to judge the distance to the needle and to the thread. But with two eyes open you can do it very easily, or at least learn easily how to do it. When you look at the object with two eyes, you automatically focus them both on the object. The closer the object the more you have to turn your eyes toward each other, and

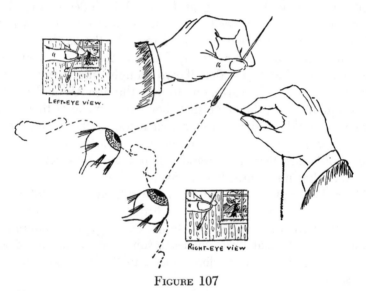

LEFT-EYE VIEW.

RIGHT-EYE VIEW

FIGURE 107

the muscular feeling arising from such adjustment gives you a pretty good idea about the distance.

Now if instead of looking with both eyes, you close first one and then the other, you will notice that the position of the object (the needle in this case) relative to the distant background (say, the window across the room) has changed. This effect is known as *parallactic displacement* and is certainly familiar to everybody; if you never heard about it, just try it out or look at Figure 107 showing the needle and window as seen by the right and the left eye. The farther away the object, the smaller will be its *parallactic displacement,* so that we can use it for measuring dis·

tances. Since *parallactic displacement* can be measured exactly in the degrees of the arc, this method is more precise than a simple judgment of the distance based on the muscular feeling in the eyeballs. But since the two eyes are set in our head only about three inches apart, they are not good for the estimate of distances beyond a few feet; in the case of more distant objects the axis of both eyes become almost parallel and the parallactic displacement becomes immeasurably small. In order to judge greater distances we should need to move our two eyes farther apart, thus increasing the angle of the parallactic displacement.

FIGURE 108

No, no surgical operation is necessary, and the trick can be done with mirrors.

In Figure 108 we see such an arrangement as used in the Navy (before the invention of radar) to measure the distance to enemy warships during battle. It is a long tube with two mirrors (*A, A'*) in front of each eye, and two other mirrors (*B, B'*) at opposite ends of the tube. Looking through such a range finder you actually see with one eye from the end *B* and with another from the end *B'*. The distance between your eyes, or the so-called optical base, becomes effectively much greater, and you can estimate much longer distances. Of course, the Navy men do not rely on just the distance-feeling given by the muscles of their eyeballs. The range finders are equipped with special gadgets

FIGURE 109

and dials measuring parallactic displacement with the utmost precision.

However these naval range finders, working perfectly even when the enemy ship is almost behind the horizon, would fail badly in any attempt to measure the distance even to such a comparatively near-by celestial body as the moon. In fact in order to notice the parallactic displacement of the moon in respect to the background of distant stars the optical base, that is, the distance between the two eyes must be made at least several hundred miles long. Of course it isn't necessary to arrange the optical system that would permit us to look with one eye from, say, Washington, and with another from New York, since all one has to do is to take two simultaneous photographs of the moon among the surrounding stars from these two cities. If you put this double picture in an ordinary stereoscope you will see the moon hanging in space in front of the stellar background. By measuring the photographs of the moon and the surrounding stars taken at the same instant in two different places on the surface of the Earth (Figure 109), astronomers have found that the parallactic displacement of the moon as it would be observed from the two opposite points of the Earth's diameter is 1°24′5″. From this it follows that the distance to the moon equals 30.14 earth-diameters, that is, 384,403 km, or 238,857 miles.

From this distance and the observed angular diameter we find that the diameter of our satellite is about one fourth of the Earth's diameter. Its total surface is only one sixteenth of the Earth's surface, about the size of the African continent.

In a similar way one can measure the distance to the sun, although, since the sun is much farther away, the measurements are considerably more difficult. Astronomers have found that this distance is 149,450,000 km (92,870,000 miles) or 385 times the distance to the moon. It is only because of this tremendous distance that the sun looks about the same size as the moon; actually it is much larger, its diameter being 109 times that of the Earth's diameter.

If the sun were a large pumpkin, the Earth would be a pea, the moon a poppy seed, and the Empire State Building in New York about as small as the smallest bacteria we can see through the microscope. It is worth while to remember here that at the time of ancient Greece, a progressive philosopher called Anaxagoras was punished with banishment and threatened with death for teaching that the sun was a ball of fire as big perhaps as the entire country of Greece.

In a similar way astronomers are able to estimate the distance
of different planets of our system. The most distant of them, discovered only quite recently and called Pluto is about forty times
farther from the sun than the Earth; to be exact, the distance is
3,668,000,000 miles.

2. THE GALAXY OF STARS

Our next jump into space will be that from the planets to the
stars, and here again the method of parallax can be used. We find,
however, that even the nearest stars are so far away that at the
most distant available observation points on the Earth (opposite
sides of the globe) they do not show any noticeable parallactic
shift in respect to the general stellar background. But we still
have a way to measure these tremendous distances. If we use the
dimensions of the Earth to measure the size of the Earth's orbit
around the sun, why don't we use this orbit to get the distances
to the stars? In other words is it not possible to notice the relative
displacements of at least some of the stars by observing them
from the opposite ends of the Earth's orbit. Of course it means
that we have to wait half a year between the two observations,
but why not?

With this idea in mind, the German astronomer Bessel started
in 1838 the comparison of the relative position of stars as observed two different nights half a year apart. First he had no
luck; the stars he picked up were evidently too far away to show
any noticeable parallactic displacement, even with the diameter
of the earth's orbit as the basis. But lo, here was the star, listed
in astronomical catalogues as 61 Cygni (61st faint star in the
constellation of Swan), which seemed to have been slightly off
its position half a year before. (Figure 110).

Another half a year passed and the star was again back in its
old place. So it was the parallactic effect after all, and Bessel was
the first man who with a yardstick stepped into the interstellar
space beyond the limits of our old planetary system.

The observed annual displacement of 61 Cygni was very small
indeed; only 0.6 angular seconds,[2] that is, the angle under which

[2] More exactly 0.″600±0.06.

you would see a man 500 miles away if you could see so far at all! But astronomical instruments are very precise, and even such angles can be measured with a high degree of accuracy. From the observed parallax, and the known diameter of the Earth's orbit, Bessel calculated that his star was 103,000,000,000,000 km away, that is, 690,000 times farther away than the sun! It is rather hard to grasp the significance of that figure. In our old example, in which the sun was a pumpkin and the Earth a pea rotating around it at a distance of 200 ft, the distance of that star would correspond to 30,000 miles!

In astronomy it is customary to speak of very large distances by giving the time they could be covered by light that travels at

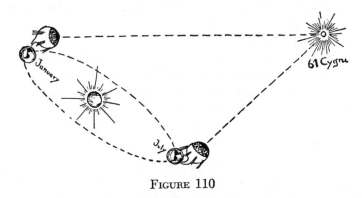

FIGURE 110

the tremendous velocity of 300,000 km per sec. It would take light only $\frac{1}{7}$ second to run around the Earth, slightly more than 1 second to come here from the moon, and about 8 minutes from the sun. From the star 61 Cygni, which is one of our nearest cosmic neighbors, the light travels to the Earth for about 11 years. If, because of some cosmic catastrophe the light from 61 Cygni were extinguished, or (what often happens to the stars) it were to explode in a sudden flash of fire, we should have to wait for 11 long years until the flash of the explosion, speeding through the interstellar space, and its last expiring ray finally brought to earth the latest cosmic news that a star had ceased to exist.

From the measured distance separating us from 61 Cygni, Bessel calculated that this star, appearing to us as a tiny luminɔus point quietly twinkling against the dark background of the night sky, is actually a giant luminous body only 30 per cent smaller and slightly less luminous than our own gorgeous sun. This was the first direct proof of the revolutionary idea first expressed by Copernicus that our sun is only one of the myriads of stars scattered at tremendous distances throughout infinite space.

Since the discovery of Bessel a great many stellar parallaxes have been measured. A few of the stars were found to be closer to us than 61 Cygni, the nearest being alpha-Centauri (the brightest star in the constellation of Centaurus), which is only 4.3 light-years away. It is very similar to our sun in its size and luminosity. Most of the stars are much farther away, so far away that even the diameter of the Earth's orbit becomes too small as the base for distance measurements.

Also the stars have been found to vary greatly in their sizes and luminosities, from shining giants such as Betelgeuse (300 light-years away), which is about 400 times larger and 3600 times brighter than our sun, to such faint dwarfs as the so-called Van Maanen's star (13 light-years away), which is smaller than our Earth (its diameter being 75 per cent that of Earth) and about 10,000 times fainter than the sun.

We come now to the important problem of counting all existing stars. There is a popular belief, to which you also probably would subscribe, that nobody can count the stars in the sky. However, as is true of so many popular beliefs, this one is also quite wrong, at least as far as the stars visible to the naked eye are concerned. In fact, the total number of stars which may thus be seen in both hemispheres is only between 6000 and 7000, and since only one half of them are above the horizon at any one time, and since the visibility of stars close to the horizon is greatly reduced by atmospheric absorption, the number of stars which are usually visible to the naked eye on a clear moonless night is only about 2000. Thus, counting diligently at the rate of say 1 star per second, you should be able to count them all in about $\frac{1}{2}$ hr!

If, however, you use a field binocular, you would be able to see some 50,000 additional stars, and a $2\frac{1}{2}$-inch telescope would reveal about 1,000,000 more. Using the famous 100-inch telescope of the Mt. Wilson observatory in California you should be able to see about half a billion stars. Counting them at the rate of 1 star per second every day from dusk to dawn, astronomers would have to spend about a century to count them all!

But, of course, nobody has ever tried to count all the stars visible through large telescopes one by one. The total number is calculated by counting the actual stars visible in a number of areas in different parts of the sky and applying the average to the total area.

More than a century ago the famous British astronomer, William Herschel, observing the stellar sky through his large self-made telescope, was struck by the fact that most of the stars that are ordinarily invisible to the naked eye appear within the faintly luminous belt cutting across the night sky and known as the Milky Way. And it is to him that the science of astronomy owes the recognition of the fact that the Milky Way is not an ordinary nebulosity or merely a belt of gas clouds spreading across space, but is actually formed from a multitude of stars that are so far away and consequently so faint that our eye cannot recognize them separately.

Using stronger and stronger telescopes we have been able to see the Milky Way as a larger and larger number of separate stars, but the main bulk of them still remains in the hazy background. It would be, however, erroneous to think that in the region of the Milky Way the stars are distributed any more densely than in any other part of the sky. It is, in fact, not the denser distribution of stars but the greater depth of stellar distribution in this direction that makes it possible to see what seems to be a larger number of stars in a given space than anywhere else in the sky. In the direction of the Milky Way the stars extend as far as the eye (strengthened by telescopes) can see, whereas in any other direction the distribution of stars does not extend to the end of visibility, and beyond them we encounter mostly the almost empty space.

Looking in the direction of the Milky Way it is as though we are looking through a deep forest where the branches of numerous trees overlap each other forming a continuous background, whereas in other directions we see patches of the empty space between the stars, as we would see the patches of the blue sky through the foliage overhead.

Thus the stellar universe, to which our sun belongs as one insignificant member, occupies a flattened area in space, extend-

FIGURE 111

An astronomer looking at the stellar system of the Milky Way reduced 100,000,000,000,000,000,000 times. The head of the astronomer is approximately in the position occupied by our sun.

ing for large distances in the plane of the Milky Way, and being comparatively thin in the direction perpendicular to it.

A more detailed study by generations of astronomers led to the conclusion that our stellar system includes about 40,000,000-000 individual stars, distributed within a lens-shaped area about

100,000 light-years in diameter and some 5000 to 10,000 light-years thick. And one result of this study comes as a slap in the face of human pride—the knowledge that our sun is not at all at the center of this giant stellar society but rather close to its outer edge.

In Figure 111 we try to convey to our readers the way this giant beehive of stars actually looks. By the way, we have not mentioned yet that in more scientific language the system of the Milky Way is known as the *Galaxy* (Latin of course!). The size of the Galaxy is here reduced by a factor of a hundred billion billions, though the number of points that represent

FIGURE 112

If we look toward the galactic center it would seem to us first that the mythical celestial road branches into two one-way traffic lanes.

separate stars are considerably fewer than forty billions, for, as one puts it, typographical reasons.

One of the most characteristic properties of the giant swarm of stars forming the galactic system is that it is in a state of rapid rotation similar to that which moves our planetary system. Just as Venus, Earth, Jupiter, and other planets move along almost circular orbits around the sun, the billions of stars forming the system of the Milky Way move around what is known as the galactic center. This center of galactic rotation is located in the direction of the constellation of Sagittarius (the Archer), and in fact if you follow the foggy shape of the Milky Way across the sky you will notice that approaching this constellation it becomes much broader, indicating that you are looking toward the central thicker part of the lens-shaped mass. (Our astronomer in Figure 111 is looking in this direction.)

What does the galactic center look like? We do not know that, since unfortunately it is screened from our sight by heavy clouds of dark interstellar material hanging in space. In fact, looking at the broadened part of the Milky Way in the region of Sagittarius[3] you would think first that the mythical celestial road branches here into two "one-way traffic lanes." But it is not an actual branching, and this impression is given simply by a dark cloud of interstellar dust and gases hanging in space right in the middle of the broadening between us and the galactic center. Thus whereas the darkness on both sides of the Milky Way is due to the background of the dark empty space, the blackness in the middle is produced by the dark opaque cloud. A few stars in the dark central patch are actually in the foreground, between us and the cloud. (Figure 112).

It is, of course, a pity that we cannot see the mysterious galactic center around which our sun is spinning, along with billions of other stars. But in a way we know how it must look, from the observation of other stellar systems or galaxies scattered through space far beyond the outermost limit of our Milky Way. It is not some supergiant star keeping in subordination all the other members of the stellar system, as the sun reigns over the family of planets. The study of the central parts of other galaxies (which we will discuss a little later) indicates that they also consist of large multitudes of stars with the only difference that here the stars are crowded much more densely than in the outlying parts to which our sun belongs. If we think of the planetary system as an autocratic state where the Sun rules the planets, the Galaxy of stars may be likened to a kind of democracy in which some members occupy influential central places while the others have to be satisfied with more humble positions on the outskirts of their society.

As said above, all the stars including our sun rotate in giant circles around the center of the galactic system. How can this be proved, how large are the radii of these stellar orbits, and how long does it take to make a complete circuit?

[3] Which can be best observed on a clear night in early summer.

All these questions were answered a few decades ago by the Dutch astronomer Oort, who applied to the system of stars known as the Milky Way observations very similar to those made by Copernicus in considering the planetary system.

Let us remember first Copernicus' argument. It had been observed by the ancients, the Babylonians, the Egyptians, and others, that the big planets like Saturn or Jupiter seemed to move

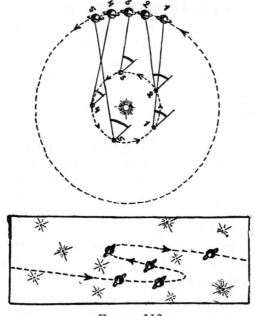

FIGURE 113

across the sky in a rather peculiar way. They seemed to proceed along an ellipse in the way the sun does, then suddenly to stop, to back, and after a second reversal of motion, to continue their way in the original direction. In the lower part of Figure 113 we show schematically such a look as described by Saturn over a period of about two years. (The period of Saturn's complete circuit is 29½ years.) Since, on account of religious prejudices that dictated the statement that our Earth is the center of the universe, all planets and the sun itself were believed to move around the Earth, the above described peculiarities of motion had to be explained by the supposition that planetary orbits have very peculiar shapes with a number of loops in them.

But Copernicus knew better, and by a stroke of genius, he explained the mysterious looping phenomenon as due to the fact that the Earth as well as all other planets move along simple circles around the Sun. This explanation of the looping effect can be easily understood after studying the schematic picture at the top of Figure 113.

The sun is in the center, the Earth (small sphere) moves along the smaller circle, and Saturn (with a ring) moves along the larger circle in the same direction as the Earth. Numbers 1, 2, 3, 4, 5 represent different positions of the Earth in the course of a year, and the corresponding positions of Saturn which, as we remember, moves much more slowly. The parts of vertical lines from the different positions of the Earth represent the direction to some fixed star. By drawing lines from the various Earth positions to the corresponding Saturn positions we see that the angle formed by the two directions (to Saturn and to the fixed star) first increases, then decreases, and then increases again. Thus the seeming phenomenon of looping does not represent any peculiarity of Saturn's motion but arises from the fact that we observe this motion from different angles on the moving Earth.

The Oort argument about the rotation of the Galaxy of stars may be understood after inspection of Figure 114. Here in the lower part of the picture we see the galactic center (with dark clouds and all!) and there are plenty of stars all around it through the entire field of the figure. The three circles represent the orbits of stars at different distances from the center, the middle circle being the orbit of our sun.

Let us consider eight stars (shown with rays to distinguish them from other points), two of which are moving along the same orbit as the sun, but one slightly ahead and one slightly behind it, the others located on somewhat larger and somewhat smaller orbits as shown in the figure. We must remember that owing to the laws of gravity (see Chapter 5) the outer stars have lower and the inner stars higher velocity than the stars on solar orbits (this is indicated in the figure by the arrows of different lengths).

How will the motion of these eight stars look if observed from the sun, or, what is of course the same, from the Earth? We are speaking here about the motion along the line of sight, which can be most conveniently observed by means of the so-called Doppler effect.[4] It is clear, first of all, that the two stars (marked D and E) that move along the same orbit and with the same speed as the sun will seem stationary to a solar (or terrestrial) observer. The same is true of the other two stars (B and G)

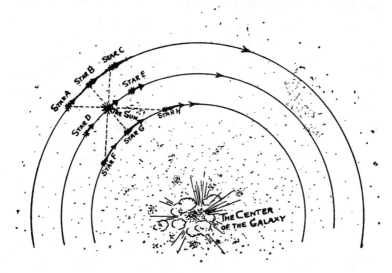

FIGURE 114

located along the radius, since they move parallel to the sun, so that there is no component of velocity along the line of sight.

Now what about the stars A and C on the outer circle? Since they both move more slowly than the sun we must conclude, as clearly seen in this picture, that the star A is lagging behind, whereas the star C is being overtaken by the sun. The distance to the star A will increase, while the distance to C will decrease, and the light coming from two stars must show respectively the red and violet Doppler effect. For the stars F and H on the inner circle the situation will be reversed, and we must have a violet Doppler effect for F and a red one for H.

[4] See the discussion of the Doppler effect on p. 330.

It is assumed that the phenomenon just described could be caused only by a circular motion of the stars, and the existence of that circular motion makes it possible for us not only to prove this assumption but also to estimate the radius of stellar orbits and the velocity of stellar motion. By collecting the observational material on the observed apparent motion of stars all over the sky, Oort was able to prove that the expected phenomenon of red and violet Doppler effect really exists, thus proving beyond any doubt the rotation of the Galaxy.

In a similar way it may be demonstrated that the effect of galactic rotation will influence the apparent velocities of stars perpendicular to the light of vision. Although this component of velocity presents much larger difficulties for exact measurement (since even very great linear velocities of distant stars correspond to extremely small angular displacements on the celestial sphere) the effect was also observed by Oort and others.

The exact measurements of the Oort effect of stellar motion now makes it possible to measure the orbits of stars and determine the period of rotation. Using this method of calculation it has been learned that the radius of the solar orbit having its center in Sagittarius is 30,000 light-years, that is, about two thirds the radius of the outermost orbit of the entire galactic system. The time necessary for the sun to move a complete circle around the galactic center is some 200 million years. It is a long time, of course, but remembering that our stellar system is about 5 billion years old, we find that during its entire life our sun with its family of planets has made about 20 complete rotations. If, following the terminology of the terrestrial year, we call the period of solar rotation "a solar year" we can say that our universe is only 20 years old. Indeed things happen slowly in the world of stars, and a solar year is quite a convenient unit for time measurements in the history of the universe!

3. *TOWARD THE LIMITS OF THE UNKNOWN*

As we have already mentioned above, our Galaxy is not the only isolated society of stars floating in the vast spaces of the universe. Telescopic studies reveal the existence, far away in space, of many other giant groups of stars very similar to that to which our sun belongs. The nearest of them, the famous Andromeda Nebula, can be seen even by the naked eye. It appears to us as a small, faint, rather elongated nebulosity. In Plates VII A and B are shown photographs, taken through the large telescope of the Mt. Wilson Observatory, of two such celestial objects. The two objects shown in these photographs are: the Nebula in Coma Berenices seen straight on the edge, and the Nebula in Ursus Major seen from the top. We notice that, as a part of the characteristic lens-shape ascribed to our Galaxy, these nebulae possess a typical spiral structure; hence the name "spiral nebulae." There are many indications that our own stellar structure is similarly a spiral, but it is very difficult to determine the shape of a structure when you are inside it. As a matter of fact, our sun is most probably located at the very end of one of the spiral arms of the "Great Nebula of the Milky Way."

For a long time astronomers did not realize that the spiral nebulae are giant stellar systems similar to our Milky Way, and confused them with ordinary diffuse nebulae like that in the constellation of Orionis, which represent the large clouds of interstellar dust floating between the stars inside our Galaxy. Later, however, it was found that these foggy spiral-shaped objects are not fog at all, but are made of separate stars, which can be seen as tiny individual points when the largest magnifications are used. But they are so far away that no parallactic measurements can indicate their actual distance.

Thus it would seem at first that we had reached the limit of our means for measuring celestial distances. But no! In science, when we come to an insuperable difficulty the delay is usually only temporary; something always happens that permits us to go still farther. In this case a quite new "measuring rod" was found by the Harvard astronomer Harlow Shapley in the so-called pulsating stars or Cepheids.[5]

There are stars and stars. While most of them glow quietly in the sky, there are a few that constantly change their luminosity from bright to dim, and from dim to bright in regularly spaced

---

[5] So called after the star β-Cephei, in which the phenomenon of pulsation was first discovered.

cycles. The giant bodies of these stars pulsate as regularly as the heart beats, and along with this pulsation goes a periodic change of their brightness.[6] The larger the star, the longer is the period of its pulsation, just as it takes a long pendulum more time to complete its swing than a short one. The really small ones (that is, small as stars go) complete their period in the course of a few hours, whereas the real giants take years and years to go through one pulsation. Now, since the bigger stars are also the brighter, there is an apparent correlation between the period of stellar pulsation, and the average brightness of the star. This relation can be established by observing the Cepheids, which are sufficiently close to us so that their distance and consequently actual brightness may be directly measured.

If now you find a pulsating star that lies beyond the limit of parallactic measurements, all you have to do is to watch the star through the telescope and observe the time consumed by its pulsation period. Knowing the period, you will know its actual brightness, and comparing it with its apparent brightness you can tell at once how far away it is. This ingenious method was successfully used by Shapley for measuring particularly large distances within the Milky Way and has been most useful in estimating the general dimensions of our stellar system.

When Shapley applied the same method to measuring the distance to several pulsating stars found imbedded in the giant body of the Andromeda Nebula, he was in for a big surprise. The distance from the Earth to these stars, which, of course, must be the same as the distance to the Andromeda Nebula itself, turned out to be 1,700,000 light-years—that is, much larger than the estimated diameter of the stellar system of Milky Way. And the size of Andromeda Nebula came out only a little smaller than the size of our entire Galaxy. The two spiral nebulae shown in our plates are still farther away and their diameters are comparable to that of the Andromeda.

This discovery dealt the death blow to the earlier assumptions that the spiral nebulae are comparatively "small things" located

[6] One must not confuse these pulsating stars with the so-called eclipsing variables, which actually represent systems of two stars rotating around each other and periodically eclipsing one another.

within our Galaxy, and established them as independent galaxies of stars very similar to our own system, the Milky Way. No astronomer would now doubt that to an observer located on some small planet circling one of the billions of stars that form the Great Andromeda Nebula, our own Milky Way would look much as the Andromeda Nebula looks to us.

The further studies of these distant stellar societies, which we owe mostly to Dr. E. Hubble, the celebrated galaxy-gazer of Mt. Wilson Observatory, reveal a great many facts of great interest and importance. It was found first of all that the galaxies, which appear more numerous through a good telescope than the ordinary stars do to the naked eye, do not all have necessarily spiral form, but present a great variety of different types. There are *spherical galaxies*, which look like regular discs with diffused

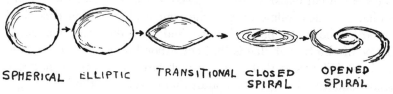

SPHERICAL   ELLIPTIC   TRANSITIONAL   CLOSED   OPENED
                                         SPIRAL   SPIRAL

FIGURE 115

Various phases of normal galactic evolution.

boundaries; there are *elliptical galaxies* with different degrees of elongation. The spirals themselves differ from each other by the "tightness with which they are wound up." There are also very peculiar shapes known as "bared spirals."

It is a fact of extreme importance that all the varieties of the observed galactic shapes can be arranged in a regular sequence (Figure 115), which presumably corresponds to different stages of the evolution of these giant stellar societies.

Although we are still far from understanding the details of galactic evolution, it seems very probable that it is due to the process of progressive contraction. It is well known that when a slowly rotating spherical body of gas undergoes a steady contraction, its speed of rotation increases, and its shape becomes that of a flattened ellipsoid. At a certain stage of contraction, when the ratio of the polar radius to the equatorial radius becomes equal to 7/10, the rotating body must assume a lenticular shape with a sharp edge running along its equator. Still further contraction keeps this lenticular shape intact, but the gases forming the rotating body begin to flow away into surrounding space all along the sharp equatorial edge, leading to the formation of a thin gaseous veil in the equatorial plane.

All the above statements have been proved mathematically by the celebrated English physicist and astronomer Sir James Jeans for a rotating gas sphere, but they can also be applied without any alterations to the giant stellar clouds that we call galaxies. In fact, we can consider such a clustering of the billions of stars as a flock of gas in which the rôle of molecules is now played by individual stars.

In comparing the theoretical calculations of Jeans with Hubble's empirical classification of galaxies, we find that these giant stellar societies follow exactly the course of evolution described by the theory. In particular we find that the most elongated shape of elliptic nebulae is that corresponding to the radius-ratio of 7/10 (E7), and that it is the first case in which we notice a sharp equatorial edge. The spirals that develop in the later stages of evolution are apparently formed from the material ejected by the rapid rotation, although up to the present we do not have a completely satisfactory explanation of why and how these spiral forms are formed and what causes the difference between the simple and the barred spirals.

Much is still to be learned from further study of the structure, motion, and stellar content in the different parts of galactic societies of stars. A very interesting result was, for example, obtained a couple of years ago by a Mt. Wilson astronomer, W. Baade, who was able to show that, whereas the central bodies (nuclei) of spiral nebulae are formed by the same type of stars as the spherical and elliptic galaxies, the arms themselves show a rather different type of stellar population. This "spiral-arm" type of stellar population differs from the population of the central region by the presence of very hot and bright stars, the so-called "Blue Giants," which are absent in the central regions as well as in the spherical and elliptic galaxies. Since, as we shall see later (Chapter XI), the Blue Giants most probably represent the most recently formed stars, it is reasonable to assume that the spiral arms are so to speak the breeding grounds for new stellar populations. One could imagine that a large part of the material ejected from the equatorial bulge of a contracting elliptic galaxy is formed by primordial gases that come out into the cold inter-galactic space and condense into the separate large lumps of matter, which through subsequent contraction become very hot and very bright.

In Chapter XI we shall return again to the problems of stellar birth and life, but now we must consider generally the distribu-tion of separate galaxies through the vastness of the universe.

We must state here, first of all that the method of distance measurements based on pulsating stars, though giving excellent results when applied to quite a number of galaxies that lie in the neighborhood of our Milky Way, fails when we proceed into the depth of space, since we soon reach distances at which no separate stars may be distinguished and the galaxies look like tiny elongated nebulosities even through the strongest telescopes. Beyond this point we can rely only on the visible size, since it is fairly well established that, unlike stars, all galaxies of a given type are of about the same size. If you know that all people are of the same height, that there are no giants or dwarfs, you can always say how far a man is from you by observing his apparent size.

Using this method for estimating distances in the far-outflung realm of galaxies, Dr. Hubble was able to prove that the galaxies are scattered more or less uniformly through space as far as the eye (fortified by the most highly powered telescope) can see. We say "more or less" because there are many cases in which the galaxies cluster in large groups containing sometimes many thousands of members, in the same way as the separate stars cluster in galaxies.

Our own galaxy, Milky Way, is apparently one member of a comparatively small group of galaxies numbering in its member-ship three spirals (including ours, and the Andromeda Nebula) and six elliptical and four irregular nebulae (two of which are Magellanic clouds).

However, save for such occasional clustering, the galaxies, as seen through the 200-inch telescope of the Mt. Palomar observatory, are scattered rather uniformly through space up to a distance of one billion light-years. The average distance between two neighboring galaxies is about 5,000,000 light-years, and the visible horizons of the universe contain about several billion individual stellar worlds!

In our old simile, in which the Empire State building was symbolized by a bacterium, the Earth by a pea, and the sun by a pumpkin, the galaxies might be represented by giant swarms of many billions of pumpkins distributed roughly within the orbit of Jupiter, separate pumpkin clusters being scattered through a spherical volume with a radius only a little smaller than the distance to the nearest star. Yes, it is very difficult to find the proper scale in cosmic distances, so that even when we scale the Earth to a pea, the size of the known universe comes out in astronomical numbers! In Figure 116 we try to give you an idea of how, step by step, astronomers have proceeded in their exploration of cosmic distance. From the Earth, to the moon, to the sun, to the stars, to distant galaxies, and toward the limits of the unknown.

We are now prepared to answer the fundamental question concerning the size of our universe. *Shall we consider the universe as extending into infinity and conclude that bigger and better telescopes will always reveal to the inquiring eye of an astronomer new and hitherto unexplored regions of space, or must we believe, on the contrary, that the universe occupies some very big but nevertheless finite volume, and is, at least in principle, explorable down to the last star?*

When we speak of the possibility that our universe is of "finite size," we do not mean, of course, that somewhere at a distance of several billion light-years the explorer of space will encounter a blank wall on which is posted the notice "No trespassing."

In fact, we have seen in Chapter III that *space can be finite without being necessarily limited by a boundary*. It can simply curve around and "close on itself," so that a hypothetical space explorer, trying to steer his rocket ship as straight as possible will describe a geodesic line in space and come back to the point from which he started.

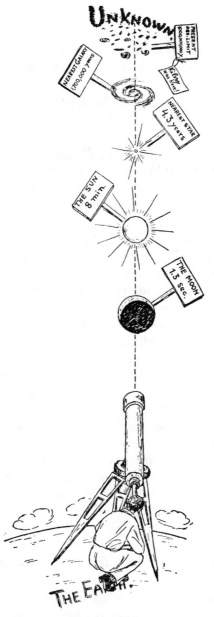

FIGURE 116

The milestones of cosmic exploration, the distances expressed in light-years.

The situation would be, of course, quite similar to an ancient Greek explorer who travels *west* from his native city of Athens, and, after a long journey, finds himself entering the eastern gates of the city.

And just as the curvature of the Earth's surface can be established without a trip around the world, simply by studying the geometry of only a comparatively small part of it, the question about the curvature of the three-dimensional space of the universe can be answered by similar measurements made within the range of available telescopes. We have seen in Chapter 5 that one must distinguish between two kinds of curvatures: the positive one corresponding to the closed space of finite volume, and the negative one corresponding to the saddle-like opened infinite space (*cf.* Figure 42). The difference between these two types of space lies in the fact that, whereas in the *closed space* the number of uniformly scattered objects falling within a given distance from the observer increases more slowly than the cube of that distance, the opposite is true in *opened space.*

In our universe the role of the "uniformly scattered objects" is played by the separate galaxies, so that all we have to do in order to solve the problem of the universal curvature is to count the number of individual galaxies located at different distances from us.

Such counting actually has been accomplished by Dr. Hubble, who has discovered that *the number of galaxies seem to increase somewhat more slowly than the cube of the distance, thus indicating the positive curvature and the finiteness of space.* It must be noticed, however, that the effect observed by Hubble is very small, becoming noticeable only near the very limit of the distance that it is possible to observe through the 100-inch Mt. Wilson telescope, and recent observations with the new 200-inch reflector on Mt. Palomar have not as yet thrown more light on this important problem.

Another point contributing to the uncertainty of the final answer concerning the finiteness of the universe lies in the fact that the distances of the faraway galaxies must be judged exclusively on the basis of their apparent luminosities (the law of inverse square). This method, which assumes that all the galaxies possess the same mean luminosity, may, however, lead to the wrong results if the luminosity of individual galaxies changes with time, thus indicating that luminosity depends on age. It must be remembered, in fact, that the most distant galaxies seen through the Mt. Palomar telescope are one billion light-years away, and are therefore seen by us in the state in which they were one billion years ago. If the galaxies become gradually fainter as they grow older (owing, perhaps, to a diminishing number of active stellar bodies as individual members die out) the conclusions reached by Hubble must be corrected. In fact, the change of galactic luminosities by only a small percentage in the course of one billion years (only about one seventh of their total age) would reverse the present conclusion that the universe is finite.

Thus we see that quite a lot of work has yet to be done before we can tell for sure whether our universe is finite or infinite.

CHAPTER XI

The Days of Creation

1. *THE BIRTH OF PLANETS*

FOR us, the people living in the seven parts of the World (counting Admiral Byrd for Antarctica) the expression "solid ground" is practically synonymous with the idea of stability and permanence. As faı as we are concerned all the familiar features of the Earth's surface, its continents and oceans, its mountains and rivers could have existed since the beginning of time. True, the data of historical geology indicate that the face of the Earth is gradually changing, that large areas of the continents may become submerged by the waters of the oceans, whereas submerged areas may come to the surface.

We also know that the old mountains are gradually being washed away by the rain, and that new mountain ridges rise from time to time as the result of tectonic activity, but all these changes are still only the changes of the solid crust of our globe.

It isn't, however, difficult to see that there must have been a time when no such solid crust existed at all, and when our Earth was a glowing globe of melted rocks. In fact, the study of the Earth's interior indicates that most of its body is still in a molten state, and that the "solid ground" of which we speak so casually is actually only a comparatively thin sheet floating on the surface of the molten magma. The simplest way to arrive at this conclusion is to remember that the temperature measured at different depths under the surface of the Earth increases at the rate of about 30° C per kilometer of depth (or 16° F per thousand feet) so that, for example, in the world's deepest mine (a gold mine in Robinson Deep, South Africa) the walls are so hot that an air-conditioning plant had to be installed to prevent the miners from being roasted alive.

At such a rate of increase, the temperature of the Earth must reach the melting point of rocks (between 1200° C and 1800° C) at a depth of only 50 km beneath the surface, that is, at less than 1 per cent of the total distance from the center. All the material farther below, forming more than 97 per cent of the Earth's body, must be in a completely molten state.

It is clear that such a situation could not have existed forever, and that we are still observing a certain stage in a process of gradual cooling that started once upon a time when the Earth was a completely molten body, and will terminate some time in the distant future with the complete solidification of the Earth's body all the way to the center. A rough estimate of the rate of cooling and growth of the solid crust indicates that the cooling process must have begun several billion years ago.

The same figure can be obtained by estimating the age of rocks forming the crust of the Earth. Although at first sight rocks exhibit no variable features, thus giving rise to the expression "unchangeable as a rock," many of them actually contain a sort of natural clock, which indicates to the experienced eye of a geologist the length of time that has passed since they solidified from their former molten state.

This age-betraying geological clock is represented by a minute amount of uranium and thorium, which are often found in various rocks taken from the surface and from different depths within the Earth. As we have seen in Chapter VII the atoms of these elements are subject to a slow spontaneous radioactive decay ending with the formation of the stable element lead.

To determine the age of a rock containing these radioactive elements we need only to measure the amount of lead that has been accumulated over the centuries as the result of radioactive decay.

In fact, as long as the material of the rock was in the molten state, the products of radioactive disintegration could have been continuously removed from the place of their origin by the process of diffusion and convection in the molten material. But as soon as the material solidified into a rock, the accumulation of lead alongside the radioactive element must have begun, and its amount can give us an exact idea of how long it was going on, in exactly the same way as the comparative numbers of empty beer cans scattered between the palms on two Pacific islands could have given to an enemy spy an idea of how long a garrison of marines had stayed on each island.

From recent surveys utilizing improved techniques for measuring precisely the accumulation in the rocks of lead isotopes and of the decay products of other unstable chemical isotopes such as rubidium-87 and potassium-40, it was estimated that the maximum age of the oldest known rocks is about four and a half billion years. Hence, we conclude that *the solid crust of the Earth must have been formed from previously molten material about five billion years ago.*

Thus we can picture the Earth five billion years ago as a completely molten spheroid, surrounded by a thick atmosphere of air, water-vapors, and probably other extremely volatile substances.

How did this hot lump of cosmic matter come into being, what kind of forces were responsible for its formation, and who supplied the material for its construction? These questions, pertaining to the origin of our Globe as well as to the origin of every other planet of our solar system, have been the basic inquiries of scientific *Cosmogony* (the theory of the origin of the universe), the riddles that have occupied the brains of astronomers for many centuries.

The first attempt to answer these questions by scientific means was made in 1749 by the celebrated French naturalist George-Louis Leclerc, Comte de Buffon, in one of the forty-four volumes of his *Natural History.* Buffon saw the origin of the planetary system as the result of a collision between the sun and a comet that came from the depth of interstellar space. His imagination painted a vivid picture of a "cométe fatale" with a long brilliant tail brushing the surface of our, at that time lonely, sun, and tearing from its giant body a number of small "drops," which were sent spinning into space by the force of the impact (Figure 117a).

A few decades later entirely different views concerning the origin of our planetary system were formulated by the famous German philosopher Immanuel Kant, who was more inclined to think that the sun made up its planetary system all by itself without the intervention of any other celestial body. Kant visualized the early state of the sun as a giant, comparatively cool, mass of gas occupying the entire volume of the present planetary system, and rotating slowly around its axis. The steady cooling of the sphere through radiation into the surrounding empty space must have led to its gradual contraction and to the corresponding increase of its rotational speed. The increasing centrifugal force

BUFFON'S COLLISION-HYPOTHESIS   KANT'S RING-HYPOTHESIS

FIGURE 117

Two schools of thought in cosmogony.

resulting from such rotation must have led to the progressive flattening of the gaseous body of the primitive sun, and resulted in the ejection of a series of gaseous rings along its extended equator (Figure 117b). Such a ring formation from the rotating masses can be demonstrated by the classical experiment performed by Plateau in which a large sphere of oil (not gaseous, as in the case of the sun) suspended within some other liquid with equal density and brought into rapid rotation by some auxiliary mechanical device begins to form rings of oil around itself when the speed of rotation exceeds a certain limit. The rings formed in this way were supposed to have broken up later and to have condensed into various planets circling at different distances around the sun.

These views were later adopted and developed by the famous French mathematician Pierre-Simon, Marquis de Laplace, who presented them to the public in his book *Exposition du système du monde*, published in 1796. Although a great mathematician, Laplace did not attempt to give mathematical treatment to these ideas, but limited himself to a semipopular qualitative discussion of the theory.

When such a mathematical treatment was first attempted sixty years later by the English physicist Clerk Maxwell, the cosmogonical views of Kant and Laplace ran into a wall of apparently insurmountable contradiction. It was, in fact, shown that if the material concentrated at present in various planets of the solar system was distributed uniformly through the entire space now occupied by it, the distribution of matter would have been so thin that the forces of gravity would have been absolutely unable to collect it into separate planets. Thus the rings thrown out from the contracting sun would forever remain rings like the ring of Saturn, which is known to be formed by innumerable small particles running on circular orbits around this planet and showing no tendency toward "coagulation" into one solid satellite.

The only escape from this difficulty would consist in the assumption that the primordial envelope of the sun contained much more matter (at least 100 times as much) than we now find in the planets, and that most of this matter fell on the sun, leaving only about 1 per cent to form planetary bodies.

Such an assumption would lead, however, to another no less serious contradiction. Indeed if so much material, which must originally have rotated with the same speed as the planets do, had fallen on the sun, it would inevitably have communicated to it an angular velocity 5000 times larger than that which it actually has. If this were the case, the sun would spin at a rate of 7 revolutions per hour instead of at 1 revolution in approximately 4 weeks.

These considerations seemed to spell death to the Kant-Laplace views, and with the eyes of astronomers turning hopefully elsewhere, Buffon's collision theory was brought back to life by the works of the American scientists T. C. Chamberlin and F. R. Moulton, and the famous English scientist Sir James Jeans. Of course, the original views of Buffon were considerably modernized by certain essential knowledge that had been gained since they were formulated. The belief that the celestial body that had collided with the sun was a comet was now discarded, for the mass of a comet was by then known to be negligibly small even when compared with the mass of the moon. And so the assaulting body was now believed rather to be another star comparable to the sun in its size and mass.

However, the regenerated collision theory, which seemed at that time to represent the only escape from the fundamental difficulties of the Kant-Laplace hypothesis, likewise found itself treading on muddy ground. It was very difficult to understand why the fragments of the sun thrown out as a result of the vigorous punch delivered by another star would move along the almost circular orbits followed by all planets, instead of describing elongated elliptical trajectories.

To save the situation it was necessary to assume that, at the time the planets were formed by the impact of the passing star, the sun was surrounded by a uniformly rotating gaseous envelope, which helped to turn the originally elongated planetary orbits into regular circles. Since no such medium is now known to exist in the region occupied by the planets, it was assumed that it was later gradually dissipated into interstellar space, and that the faint luminosity known as *Zodiacal Light*, which at present extends from the sun in the plane of ecliptics, is all that is left from that past glory. But this picture, representing a kind of hybrid between the Kant-Laplace assumption of the original gaseous envelope of the sun and Buffon's collision hypothesis was very unsatisfactory. However, as the proverb says, one must choose the lesser of two evils, and the collision hypothesis of the origin of the planetary system was accepted as the correct one, being used until very recently in all scientific treatises, textbooks, and popular literature (including the author's two books *The*

Birth and Death of the Sun, 1940, and *Biography of the Earth*, revised edition 1959; first published 1941).

It was only in the fall of 1943 that the young German physicist C. Weizsäcker cut through the Gordian Knot of the planetary theory. Using the new information collected by recent astrophysical research, he was able to show that all the old objections against the Kant-Laplace hypothesis can be easily removed, and that, proceeding along these lines, one can build a detailed theory of the origin of planets, explaining many important features of the planetary system that had not even been touched by any of the old theories.

The main point of Weizsäcker's work lies in the fact that during the last couple of decades astrophysicists have completely changed their minds about the chemical constitution of matter in the universe. It was generally believed before that the sun and all other stars were formed by the same percentage of chemical elements as those that we have learned from our Earth. Geochemical analysis teaches us that the body of the Earth is made up chiefly of oxygen (in the form of various oxides), silicon, iron, and smaller quantities of other heavier elements. Light gases such as hydrogen and helium (along with other so-called rare gases such as neon, argon, etc.) are present on the Earth in very small quantities.[1]

In the absence of any better evidence, astronomers had assumed that these gases were also very rare in the bodies of the sun and the other stars. However, the more detailed theoretical study of stellar structure led the Danish astrophysicist B. Stromgren to the conclusion that such an assumption is quite incorrect and that, in fact, at least 35 per cent of the material of our sun must be pure hydrogen. Later this estimate was increased to above 50 per cent, and it was also found that a considerable percentage of the other solar constituents is pure helium. Both the theoretical studies of the solar interior (which recently culminated in the important work of M. Schwartzschild), and the more elaborate spectroscopic analysis of its surface, led astro-

---

[1] Hydrogen is found on our planet mostly in its union with oxygen in water. But everybody knows that although water covers three quarters of the Earth's surface the total water mass is very small compared with the mass of the entire body of the Earth.

physicists to a striking conclusion that: *the common chemical elements that form the body of the Earth constitute only about 1 per cent of the solar mass, the rest being almost evenly divided between hydrogen and helium with a slight preponderance of the former.* Apparently this analysis also fits the constitution of the other stars.

Further, it is now known that *interstellar space is not quite empty,* but is filled by a mixture of gas and fine dust with a *mean density of about 1 mg matter in 1,000,000 cu miles space,* and this diffuse, highly rarified material apparently has the same chemical constitution as have the sun and the other stars.

In spite of its incredibly low density the presence of this interstellar material can be easily proved, since it produces noticeable selective absorption of the light from stars so distant that it has to travel for hundreds of thousands of light-years through space before entering into our telescopes. The intensity and location of these "interstellar absorption lines" permits us to obtain good estimates of the density of that diffuse material and also to show that it consists almost exclusively of hydrogen and probably helium. In fact, the dust, formed by small particles (about 0.001 mm in diameter) of various "terrestrial" materials, constitutes not more than 1 per cent of its total mass.

To return to the basic idea of Weizsäcker's theory, we may say that this new knowledge concerning the chemical constitution of matter in the universe, plays directly into the hand of the Kant-Laplace hypothesis. In fact, if the primordial gaseous envelope of the sun was originally formed from such material, *only a small portion of it, representing heavier terrestrial elements, could have been used to build our Earth and other planets.* The rest of it, represented by noncondensible hydrogen and helium gases, must have been somehow removed, either by falling into the sun or by being dispersed into surrounding interstellar space. Since the first possibility would result, as it was explained above, in much too rapid axial rotation of the sun, we have to accept another alternative, namely, that the gaseous "excess-material" was dispersed into space soon after the planets were formed from the "terrestrial" compound.

This brings us to the following picture of the formation of the

planetary system. When our sun was first formed by the condensation of interstellar matter (see the next section) a large part of
it, probably about a hundred times the present combined mass
of planets, remained on the outside forming a giant rotating envelope. (The reason for such behavior can easily be found in the
differences between the rotational states of various parts of
interstellar gas condensing into the primitive sun.) This rapidly
rotating envelope should be visualized as consisting of *noncondensible gases* (hydrogen, helium, and a smaller amount of other
gases) and *dust-particles* of various terrestrial materials (such as
iron oxides, silicon compounds, water droplets and ice crystals)
which were floating inside the gas and carried along by

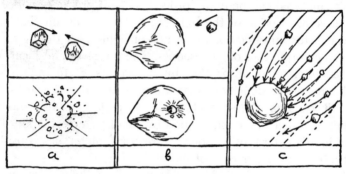

FIGURE 118

its rotational motion. The formation of big lumps of "terrestrial"
material, which we now call planets must have taken place as the
result of collisions between dust particles and their gradual aggregation into larger and larger bodies. In Figure 118 we illustrate the results of such mutual collisions which must have taken
place at velocities comparable to that of meteorites.

One must conclude, on the basis of logical reasoning, that at
such velocities the collision of two particles of about equal mass
would result in their mutual pulverization (Figure 118a), a process leading not to the growth but rather to the destruction of
larger lumps of matter. On the other hand, when a small particle
collides with a much larger one (Figure 118b) it seems evident
that it would bury itself in the body of the latter, thus forming
a new, somewhat larger mass.

Obviously these two processes would result in the gradual disappearance of smaller particles and the aggregation of their material into larger bodies. In the later stages the process will be accelerated due to the fact that the larger lumps of matter will attract gravitationally the smaller particles passing by and add them to their own growing bodies. This is illustrated in Figure 118c, which shows that in this case the capture-effectiveness of massive lumps of matter becomes considerably larger.

Weiszäcker was able to show that *the fine dust originally scattered through the entire region now occupied by the planetary*

Name of the planet	Distance from the sun in terms of earth's distance from the sun	The ratio of the distance of each planet from the sun, to the distance from the sun of the planet listed above it
Mercury	0.387	
Venus	0.723	1.86
Earth	1.000	1.38
Mars	1.524	1.52
Planetoids	about 2.7	1.77
Jupiter	5.203	1.92
Saturn	9.539	1.83
Uranus	19.191	2.001
Neptune	30.07	1.56
Pluto	39.52	1.31

system must have been aggregated into a few big lumps to form the planets, within a period of about a hundred million years.

As long as the planets were growing by the accretion of variously sized pieces of cosmic matter on their way around the sun, the constant bombardment of their surfaces by fresh building material must have kept their bodies very hot. As soon, however, as the supply of stellar dust, pebbles, and larger rocks was exhausted, thus stopping the process of further growth, the radiation into interstellar space must have rapidly cooled the outer layers of the newly formed celestial bodies, and led to the formation of the solid crust, which is even now growing thicker and thicker, as the slow internal cooling continues.

The next important point to be attacked by any theory of planetary origin is the explanation of the peculiar rule (known as the *Titus-Bode* rule) that governs the distances of different planets from the sun. In the table on page 307, these distances are listed for nine planets of the solar system, as well as for the belt of *planetoids*, which apparently corresponds to an exceptional case where separate pieces did not succeed in collecting themselves into a single big lump.

The figures in the last column are of especial interest. In spite of some variations, it is evident that none are very far from the

Name of satellite	Distance in terms of Saturn's radius	The ratio of increase in two successive distances
Mimas	3.11	
Enceladus	3.99	1.28
Tethys	4.94	1.24
Dione	6.33	1.28
Rhea	8.84	1.39
Titan	20.48	2.31
Hyperion	24.82	1.21
Japetus	59.68	2.40
Phoebe	216.8	3.63

numeral 2, which permits us to formulate the approximate rule: *the radius of each planetary orbit is roughly twice as large as that of the orbit nearest to it in the direction of the sun.*

It is interesting to notice that a similar rule holds also for the satellites of individual planets, a fact that can be demonstrated, for example, by the above table giving the relative distances of nine satellites of Saturn.

As in the case of the planets themselves, we encounter here quite large deviations (especially for Phoebe!) but again there is hardly any doubt that there is a definite trend for regularity of the same type.

How can we explain the fact that the aggregation process that took place in the original dust cloud surrounding the sun did not result in the first place in just one big planet, and why the several big lumps were formed at these particular distances from the sun?

To answer this question we have to undertake a somewhat more detailed survey of motions that took place in the original dust cloud. We must remember first of all that every material body—whether it is a tiny dust particle, a small meteorite, or a big planet—that moves around the sun under the Newtonian law of attraction is bound to describe an elliptical orbit with the sun in the focus. If the material forming the planets was formerly

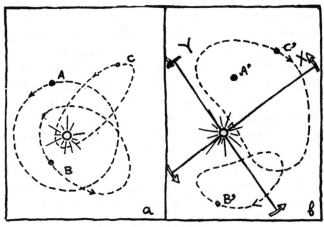

FIGURE 119

Circular and elliptic motion as viewed from a resting (*a*) and a rotating (*b*) co-ordinate system

in the form of separate particles, say, 0.0001 cm. in diameter,[2] there must have been some 10^{45} particles moving along the elliptical orbits of all various sizes and elongations. It is clear that, in such heavy traffic, numerous collisions must have taken place between the individual particles, and that, as the result of such collisions, the motion of the entire swarm must have become to a certain extent organized. In fact, it is not difficult to understand that such collisions served either to pulverize the "traffic violators" or to force them to "detour" into less crowded "traffic

---

[2] The approximate size of the dust particles forming the interstellar material.

lanes." What are the laws that would govern such "organized" or at least partially organized "traffic"?

To make the first approach to the problem, let us select a group of particles all of which had the *same rotation-period* around the sun. Some of them were moving along the circular orbit of a corresponding radius, whereas others were describing various more or less elongated elliptical orbits (Figure 119*a*). Let us now try to describe the motion of these various particles from the point of view of a co-ordinate system (*X, Y*) rotating around the center of the sun with the same period as the particles.

It is clear first of all that, from the point of view of such a rotating co-ordinate system, the particle that was moving along a circular orbit (*A*) would appear to be completely at rest at a certain point *A'*. A particle *B* that was moving around the sun following an elliptical trajectory comes closer and farther away from the sun; and its angular velocity around the center is larger in the first case and smaller in the second; thus, it will sometimes run ahead of the uniformly rotating co-ordinate system (*X, Y*), and sometimes will lag behind. It is not difficult to see that, from the point of view of this system, the particle will be found to describe *a closed bean-shaped trajectory* marked *B'* in Figure 119. Still another particle *C,* which was moving along a more elongated ellipse, will be seen in the system (*X, Y*) as describing a similar but somewhat larger bean-shaped trajectory *C'*.

It is clear now that, if we want to arrange the motion of the entire swarm of particles so that they never collide with one another, *it must be done in such a way that the bean-shaped trajectories described by these particles in the uniformly rotated co-ordinate system (X, Y) do not intersect.*

Remembering that the particles having *common rotation periods* around the sun keep the same *average distance* from it, we find that the nonintersecting pattern of their trajectories in the system (*X, Y*) must look like a "bean necklace" surrounding the sun.

The aim of the above analysis, which may be a bit too hard on the reader, but which represents in principle a fairly simple procedure, is to show *the nonintersecting traffic-rules pattern* for individual groups of particles moving at the same mean distance from the sun and possessing therefore the same period of rotation. Since in the original dust cloud surrounding the primitive sun we should expect to encounter all different mean distances and correspondingly all different rotation periods, the actual situation must have been more complicated. Instead of just one "bean necklace" there must have been a large number of such "necklaces" rotating in respect to one another with various speeds. By careful analysis of the situation, Weizsäcker was able to show

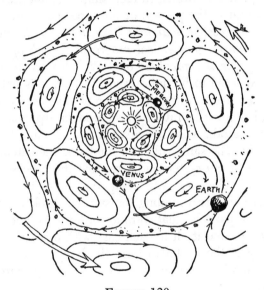

FIGURE 120

Dust-traffic lanes in the original solar envelope.

that for the stability of such a system it is necessary that each separate "necklace" should contain five separate whirlpool systems so that the entire picture of motion must have looked very much like Figure 120. Such an arrangement would assure "safe traffic" within each individual ring, but, since these rings rotated with different periods, there must have been "traffic accidents" where one ring touched another. The large number of mutual collisions taking place in these boundary regions between the particles belonging to one ring and those belonging to neighboring rings must have been responsible for the aggregation process and for the growth of larger and larger lumps of matter at these particular distances from the sun. Thus, through a gradual thinning process within each ring, and through the accumulation of matter at the boundary regions between them, the planets were finally formed.

The above described picture of the formation of the planetary system gives us a simple explanation of the old rule governing the radii of planetary orbits. In fact, simple geometrical considerations show that in the pattern of the type shown in Figure 120, *the radii of successive boundary lines between the neighboring rings form a simple geometrical progression, each of them being twice as large as the previous one.* We also see why this rule cannot be expected to be quite exact. In fact, it is not the result of some *strict law* governing the motion of particles in the original dust cloud, but must be rather considered as expressing a certain *tendency* in the otherwise irregular process of dust traffic.

The fact that the same rule also holds for the satellites of different planets of our system indicates that the process of satellite formation took place roughly along the same lines. When the original dust cloud surrounding the sun was broken up into separate groups of particles that were to form the individual planets, the process repeated itself in each case with most of the material concentrating in the center to form the body of the planet, and the rest of it circling around condensing gradually into a number of satellites.

With all our discussion of mutual collisions and the growth of dust particles, we have forgotten to tell what happened to the gaseous part of the primordial solar envelope that, as may be remembered, constituted originally about 99 per cent of its entire mass. The answer to this question is a comparatively simple one.

While the dust particles were colliding, forming larger and larger lumps of matter, the gases that were unable to participate in that process were gradually dissipating into interstellar space. It can be shown by comparatively simple calculations that the time necessary for such dissipation was about 100,000,000 years, that is, about the same as the period of planetary growth. Thus by the time the planets were finally formed, most of the hydrogen and helium that had formed the original solar envelope must have escaped from the solar system, leaving only the negligibly small traces referred to above as Zodiacal Light.

One important consequence of the Weizsäcker theory lies in the conclusion that *the formation of the planetary system was not an exceptional event, but one that must have taken place in the formation of practically all of the stars.* This statement stands in sharp contrast with the conclusions of the collision theory, which considered the process by which the planets were formed as very exceptional in cosmic history. In fact, it was calculated that stellar collisions that were supposed to give rise to planetary systems are extremely rare events, and that among 40,000,000,-000 stars forming our stellar system of the Milky Way, only a few such collisions could have taken place during several billion years of its existence.

If, as it appears now, *each star possesses a system of planets,* there must be millions of planets within our galaxy alone, the physical conditions on which are almost identical with those on our Earth. And it would be at least strange if life—even in its highest forms—had failed to develop in these "inhabitable" worlds.

In fact, as we have seen in Chapter IX, the simplest forms of life, such as different kinds of viruses, actually are merely rather complicated molecules composed mainly of carbon, hydrogen, oxygen, and nitrogen atoms. Since these elements must be present in sufficient abundance on the surface of any newly formed planet, we must believe that sooner or later after the formation of the solid crust of earth and the precipitation of atmospheric vapors forming the extensive water reservoirs, a few molecules of such type must have appeared, owing to an accidental combination of the necessary atoms in the necessary order. To be sure, the complexity of living molecules makes the probability of their accidental formation extremely small, and we can compare it with the probability of putting together a jigsaw puzzle by simply shaking the separate pieces in their box with the hope that they will accidentally arrange themselves in the proper way. But on the other hand we must not forget that there were an immense number of atoms continuously colliding with one another, and also a lot of time in which to achieve the necessary result. The fact that life appeared on our Earth rather soon after the formation of the crust indicates that, improbable as it seems, the accidental formation of a complex organic molecule required probably only a few hundred million years. Once the simplest forms of life appeared on the surface of the newly formed planet, the process of organic reproduction, and the gradual evolution would lead to the formation of more and more complicated forms of living organisms.[3] There is no telling whether the evolution of life on different "inhabitable" planets takes the same track as it did on our Earth. The study of life in different worlds would contribute essentially to our understanding of the evolutionary process.

But whereas we may be able to study the forms of life that may have developed on Mars and Venus (the best "inhabitable" planets of the solar system) in the not too distant future by means of an adventuresome trip to these planets on a "nuclear-power propelled space ship," the question about the possible existence and the forms of life in other stellar worlds hundreds and thousands of light-years away, will probably remain forever an unsolvable problem of science.

2. THE PRIVATE LIFE OF THE STARS

Having a more or less complete picture of how the individual stars give birth to their families of planets, we may now ask ourselves about the stars themselves.

What is the life history of a star? What are the details of its birth, through what changes does it go during its long life, and what is its ultimate end?

We can start studying this question by looking first at our own sun, which is a rather typical member among the billions of stars forming the system of the Milky Way. We know, first of all, that our sun is a rather old star, since according to the data of paleontology it has been shining with unchanged intensity for a

---

[3] More detailed discussion of the origin and evolution of life on our planet can be found in the author's book *Biography of the Earth* (New York, The Viking Press, rev. ed. 1959; first published 1941).

few billion years supporting the development of life on the Earth. No ordinary source could supply so much energy for such a long period of time, and the problem of solar radiation remained one of the most puzzling riddles of science until the discovery of radioactive transformations and the artificial transformation of

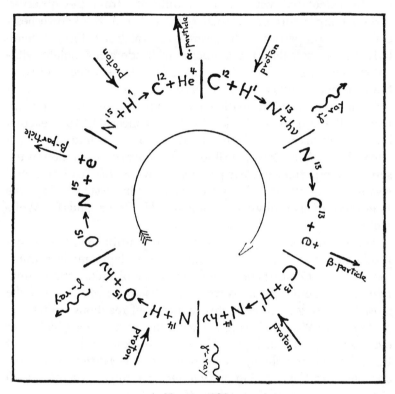

The cyclic nuclear reaction chain responsible for the energy generation in the sun.

elements revealed to us tremendous sources of energy hidden in the depths of atomic nuclei. We have already seen in Chapter 7 that practically every chemical element represents an alchemical fuel with a potentially tremendous output of energy, and that this energy can be liberated by heating up these materials to millions of degrees.

Whereas such high temperatures are practically unattainable in terrestrial laboratories, they are rather common in the stellar world. In the sun, for example, the temperature, which is only 6000° C at the surface, increases gradually inward reaching in the center the tremendous value of 20 million degrees. This figure can be calculated without much difficulty from the observed surface temperature of the solar body and from the known heat-conducting properties of the gases from which it is formed. Similarly we can calculate the temperature inside a hot potato without cutting it, if we know how hot it is on the surface, and what the heat conductivity of its material is.

Combining this information concerning the central temperature of the sun, with the known facts concerning the reaction rates of various nuclear transformations, one can find out which particular reaction is responsible for the energy production in the sun. This important nuclear process, known as the "carbon-cycle" was found simultaneously by two nuclear physicists who became interested in astrophysical problems: H. Bethe and C. Weizsäcker.

The thermonuclear process mainly responsible for the energy production of the sun is not limited to a single nuclear transformation, but consists of a whole sequence of linked transformations that together form, as we say, *a reaction chain.* One of the most interesting features of this sequence of reactions is that it is *a closed circular chain,* returning us to our starting point after every six steps. From Figure 121 which represents the scheme of this solar reaction chain, we see that *the main participants of the sequence are the nuclei of carbon and of nitrogen, together with the thermal protons with which they collide.*

Starting, for instance, with ordinary carbon (C^{12}), we see that the result of a collision with a proton is the formation of the lighter isotope of nitrogen (N^{13}), and the liberation of some subatomic energy in the form of a γ-ray. This particular reaction is well known to nuclear physicists, and has also been obtained under laboratory conditions by the use of artificially accelerated high-energy protons. The nucleus of N^{13}, being unstable, adjusts itself by emitting a positive electron, or positive β-particle, and becoming the stable nucleus of the heavier carbon isotope (C^{13}), which is known to be present in small quantities in ordinary coal. Being struck by another thermal proton, this carbon isotope is transformed into ordinary nitrogen (N^{14}), with additional intense gamma radiation. Now the nucleus of N^{14} (from which we could just as easily have begun our description of the cycle) collides with still another (third) thermal proton and gives rise to an unstable oxygen isotope (O^{15}), which very rapidly goes over to the stable N^{15} through the emission of a positive electron. Finally, N^{15}, receiving in its interior a fourth proton, splits into two unequal parts, one of which is the C^{12} nucleus with which we began and the other of which is a helium nucleus, or α-particle.

Thus we see that *the nuclei of carbon and nitrogen in our circular reaction chain are forever being regenerated, and act only as catalysts*, as chemists would say. The net result of the reaction chain is the formation of one helium nucleus from the four protons that have successively entered the cycle; and we may therefore describe the whole process as *the transformation of hydrogen into helium as induced by high temperatures and aided by the catalytic action of carbon and nitrogen.*

Bethe was able to show that the energy liberation of his reaction chain at the temperature of 20 million degrees *coincides with the actual amount of energy radiated by our sun.* Since all other possible reactions lead to results inconsistent with the astrophysical evidence, it should be definitely accepted that *the carbon-nitrogen cycle represents the process mainly responsible for solar energy generation.* It should also be noted here that at the interior temperature of the sun the complete cycle shown in Figure 121 requires about 5 million years, so that at the end of this period each carbon (or nitrogen) nucleus that originally entered the reaction will emerge from it again as fresh and untouched as it was to start with.

In view of the basic part played in this process by carbon, there is something to be said after all for the primitive view that the Sun's heat came from coal; only we know now that the "coal," instead of being a real fuel, plays rather the role of the legendary phoenix.

It must be particularly noticed here that whereas the rate of energy-producing reaction in the sun depends essentially on the temperature and density obtaining in its central regions, it must also depend to some extent on the content of hydrogen, carbon, and nitrogen in the material forming the solar body. This deduction immediately suggests a method by which we may analyze the constitution of solar gases by adjusting the concentrations of the reactants involved (i.e. the reacting substances) so as to fit exactly the observed luminosity of the sun. Calculations based on this method were made quite recently by M. Schwartzschild, with the resulting discovery that *over one half of solar matter is formed by pure hydrogen, somewhat less than one half by pure helium, and only a very small residue by all other elements.*

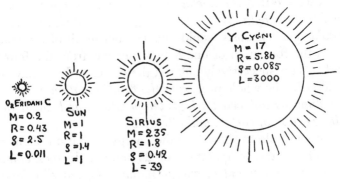

FIGURE 122

The main sequence of stars.

The explanation of energy production in the sun can easily be extended to most of the other stars, with the conclusion that the stars with different masses have different central temperatures, and consequently different rates of energy production. Thus the star known as O_2 Eridani C is about five times lighter than the sun and correspondingly shines with an intensity that is only about 1 per cent that of the sun. On the other hand X Canis Majoris A, commonly known as Sirius, is about two and a half times heavier than the sun and forty times more luminous. There are also such giant stars as, for example, Y 380 Cygni, which is about forty times heavier and several hundred thousand times brighter than the sun. In all these cases, the relation between the larger stellar mass and its much higher luminosity can be explained very satisfactorily by the increasing rate of "carbon-cycle" reaction caused by the higher central temperature. Following this so-called "Main Sequence" of stars we also find that the increasing mass results in an increasing stellar radius (from 0.43 sun radius for O_2 Eridani C to 29 sun radii for Y 380 Cygni) and its decreasing mean density (from 2.5 for O_2 Eridani C, through 1.4 for the sun, to 0.002 for Y 380 Cygni). Some data about the

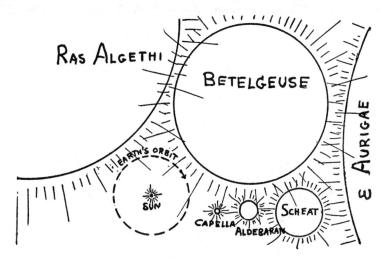

FIGURE 123

Giant and supergiant stars as compared with the size of our planetary system.

stars of the main sequence are collected in the diagram shown in Figure 122.

Apart from the "normal" stars, whose radius, density and luminosity are determined by their masses, astronomers find in the sky some stellar types that definitely fall out of this simple regularity.

First of all there are the so-called "red giant" and "supergiant" stars, which, although they have the same quantity of matter as the "normal" stars of the same luminosity, possess, however, much larger linear dimensions. In Figure 123 we give a schematic picture of this abnormal group of stars, which include such famous names as *Capella, Scheat, Aldebaran, Betelgeuse, Ras Algethi,* and *E Aurigae.*

Apparently the bodies of these stars were blown up to almost incredibly large dimensions by internal forces that we cannot yet explain, causing their mean densities to fall well below the densities of any normal stars.

In contrast to these "swollen up" stars we have another group of stars that are shrunk to very small diameters. One of the stars of this class, known as "white dwarfs,"[4] is shown in Figure 124

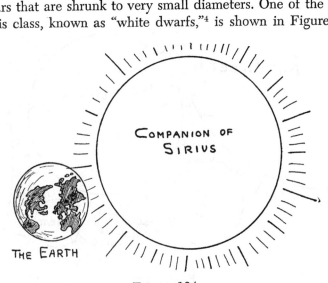

FIGURE 124

White dwarf stars as compared with the Earth.

with a diagram of the Earth for comparison. The "Companion of Sirius" is composed of a mass almost equal to that of the sun, is only three times larger than the Earth; its mean density must be about 500,000 greater than that of water! There is hardly any doubt that the white dwarf stars represent the late stages of stellar evolution corresponding to the phase in which the star has consumed all its available hydrogen fuel.

[4] The origin of the terms "red giants" and "white dwarfs" lies in the relation of their luminosity to their surfaces. Since the rarified stars have very large surfaces to radiate the energy produced in their interiors their surface temperatures are comparatively low, giving to them red coloring. The surface of the highly condensed stars, on the other hand, must be necessarily very hot, or white hot.

As we have seen above, the life sources of stars lie in the al-chemical reaction slowly transforming hydrogen into helium. Since in a young star, which was just formed by the condensation of diffused interstellar material, the content of hydrogen exceeds 50 per cent of its entire mass, we can expect that the stellar life spans are extremely long. Thus, for example, one calculates from the observed luminosity of our sun that it consumes about 660 million tons of hydrogen per second. Since the total mass of the sun is 2×10^{27} tons, half of it being hydrogen, we find that the life span of the sun must be $15 \cdot 10^{18}$ seconds or about 50 billion years! Remembering that our sun is now only about 3 or 4 billion years old,[5] we see that it must still be considered to be very young, and will continue to shine with approximately the present intensity for billions of years to come.

But the more massive, and therefore more luminous, stars spend their original hydrogen supply at a much higher rate. Thus for example Sirius being 2.3 times heavier than the sun, and therefore containing originally 2.3 times more hydrogen fuel, is 39 times more luminous than the sun. Spending 39 times as much fuel in a given time than the sun, and having an original supply only 2.3 times as great, Sirius would use it all in only 3,000,000,-000 years. In still brighter stars, such as for example Y Cygni (17 times the mass of the sun, and 30,000 times as luminous), the original hydrogen supply would not last for more than 100,000,-000 years.

What happens to a star when its hydrogen supply is finally exhausted?

Since the nuclear energy source that was supporting the star more or less *in status quo* during its long life is gone, the body of the star must begin to contract, thus going through successive stages of greater and greater density.

Astronomical observations reveal the existence of a large number of such *"shrunken stars"* the mean density of which exceeds the density of water by a factor of several hundred thousand. These stars are still very hot and owing to their high surface

[5] Since according to Weizsäcker's theory, the sun must have been formed not much before the formation of the planetary system, and since the estimated age of our earth is of that order of magnitude.

temperature shine with a brilliant white light representing a sharp contrast to the ordinary yellowish or reddish stars of the main sequence. Since, however, these stars are very small in size their total luminosity is rather low, thousands of times lower than that of the sun. Astronomers call these late stages of stellar evolution "white dwarfs," the latter term being used both in the sense of geometrical dimensions as well as in the sense of total luminosity. As time goes on the white-hot bodies of the white dwarfs will gradually lose their brilliance and they will finally turn into the "black dwarfs," the large cold masses of matter inaccessible to ordinary astronomical observations.

It must however be noticed here that the process of shrinking and gradual cooling of the old stars which have used up all their vital hydrogen fuel does not always proceed in a quiet and orderly way, and that, walking their "last mile," these dying stars are often subject to titanic convulsions as if revolting against their fate.

These catastrophic events, known as *novae and supernovae-explosions*, represent one of the most exciting topics of stellar studies. Within a few days, a star, which had not seemed before to differ much from any other star in the sky, increases its luminosity by a factor of several hundred thousand and its surface becomes evidently extremely hot. The study of the changes in the spectrum accompanying this sudden increase of luminosity indicates that the body of the star is rapidly swelling up, and that its outer layers are expanding with the velocity of about 2000 km per second. The increase of luminosity is, however, only temporary, and, after passing through the maximum, the star begins slowly to settle down. It takes usually about a year before the luminosity of the exploded star returns to its original value, though small variations of stellar radiation have been observed after considerably longer time intervals. Although the luminosity of the star becomes normal again, one cannot say the same about its other properties. A part of the stellar atmosphere, participating in the rapid expansion during the explosion phase, continues its outward motion, and the star is surrounded by a luminous gas shell of gradually increasing diameter. The evidence concerning permanent changes of the star proper is as yet very indecisive, as there is only one case in which the spectrum of the star was photographed before the explosion (Nova Aurigae 1918). But even this photograph is seemingly so imperfect that the conclusion concerning surface temperature and the radius of the pre-nova stage must be considered as very uncertain.

Somewhat better evidence concerning the result of the explosion in the body of the star can be obtained from the observations of the so-called supernovae explosions. These vast stellar explosions, which happen in our stellar system only once in several centuries (in contrast to ordinary novae, which appear at the rate of about 40 per year), exceed the luminosity of ordinary novae by a factor of several thousand. During the maximum, the light emitted by such an exploding star is comparable to the light emitted by the entire stellar system. The star observed by Tycho Brahe in 1572 and visible in bright daylight, the star registered by Chinese astronomers in the year 1054, and probably the Star of Bethlehem represent typical examples of such supernovae within our stellar system, the Milky Way.

The first extragalactic supernova was observed in 1885 in the neighboring stellar system known as The Great Andromeda Nebula, its luminosity exceeding by a factor of one thousand the luminosities of all other novae ever seen in this system. In spite of the comparative rarity of these vast explosions, the study of their properties has made considerable progress in recent years owing to observations of Baade and Zwicky, who were the first to recognize the great difference between the two types of explosions and began the systematic study of supernovae appearing in various distant stellar systems.

In spite of the tremendous difference in luminosity, the phenomena of supernovae explosions show many features similar to those of the ordinary novae. The rapid rise of luminosity and its subsequent slow decrease in both cases are represented (apart from the scale) by practically identical curves. As is true for ordinary novae, a supernova explosion gives rise to a rapidly expanding gas shell, which, however, takes a considerably larger fraction of the stellar mass. In fact, whereas the gas shells emitted by novae become thinner and thinner and dissolve themselves rapidly in the surrounding space, the gas masses emitted by supernovae form extensive luminous nebulae involving the place of explosion. It can be, for example, considered as definitely established that the so-called "Crab Nebula," seen at the place of the supernova of the year 1054, was formed by gases expelled during that explosion (see Plate VIII).

In the case of this particular supernova we also have some evidence concerning the star remaining after the explosion. In fact, in the very center of the Crab Nebula, observations show the presence of a faint star which, according to its observed properties, must be classified as a very dense white dwarf.

All this indicates that the physical processes of supernovae explosion must be analogous to those of the ordinary novae, although everything is happening on a much larger scale.

Assuming the "collapse theory" of novae and supernovae, we must first of all ask ourselves about the causes that could lead to such a rapid contraction of the entire stellar body. It is well established at present that the stars represent giant masses of hot gas, and that in the state of equilibrium the body of the star is supported entirely by the high gas pressure of the hot material in its interior. As long as the "carbon cycle" described above is proceeding in the center of the star, the energy radiated from the surface is being replenished by subatomic energy produced in the interior, and the state of the star changes but very little. As soon, however, as the hydrogen content is completely exhausted, no more subatomic energy is available and the star must begin to contract, thus turning into radiation its potential energy of gravity. The process of such gravitational contraction will, however, go very slowly, since, because of the high opacity of stellar material, the heat transport from the interior to the surface is very slow. It can be estimated, for example, that in order to contract to half of its present radius, our sun would require more than ten million years. Any attempt to contract faster than that would immediately result in the liberation of additional gravitational energy, which would increase the temperature and gas pressure in the interior and slow down the contraction. It can be seen from the above considerations that the only way to accelerate the contraction of a star and to turn it into a rapid collapse as observed in the case of novae and supernovae, would be to devise some mechanism that would remove from the interior the energy liberated in contraction. If, for example, the opacity of stellar matter could be reduced by a factor of several billions, the contraction would be accelerated in the same proportion, and a contracting star would collapse within a few days. This possibility is, however, quite excluded, since the present theory of radiation definitely shows that the opacity of stellar matter is quite definitely a function of its density and temperature, and can hardly be reduced even by so much as a factor of ten or a hundred.

It was recently proposed by the author and his colleague, Dr. Schenberg, that the real cause of stellar collapses is due to the

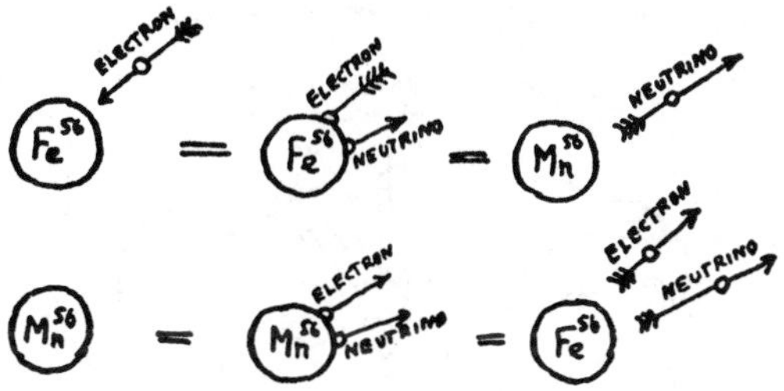

FIGURE 125

The Urca process in iron nucleus leading to the unlimited formation of neutrinos.

mass formation of *neutrinos*, those tiny nuclear particles that are discussed in detail in Chapter 7 of this book. It is clear from the description of the neutrino that it is just the right agent to remove the surplus energy from the interior of a contracting star, since the entire body of the star is just as transparent for neutrinos as a windowpane is for ordinary light. It remains to be seen whether the neutrinos will be produced, and produced in sufficiently large numbers in the hot interior of a contracting star.

The reactions that must be necessarily accompanied by emission of neutrinos consist in the capture of fast-moving electrons by the nuclei of various elements. When a fast electron penetrates inside the atomic nucleus, a high-energy neutrino is im-

mediately emitted, and the electron is retained, transforming the
original nucleus into an unstable nucleus of the same atomic
weight. Being unstable, this newly formed nucleus can exist
only a definite period of time, and subsequently decays, emitting
its electron in the company of another neutrino. Then the process
begins again from the beginning, and leads to a new neutrino
emission. . . . (Figure 125).

If the temperature and density are high enough, as they are in
the interior of contracting stars, the energy losses through

FIGURE 126

An early and a late stage of a supernova explosion.

neutrino emission will be tremendously high. Thus, for example,
the capture and re-emission of electrons by the neuclei of iron
atoms will transform into neutrino energy as much as 10^{11} ergs
per gram per second. In the case of oxygen (where the unstable
product is radioactive nitrogen with the decay period of 9
seconds) the star can lose even as much as 10^{17} ergs per second
per gram of its material. The energy losses in this latter case are
so high that the complete collapse of the star takes place in only
twenty-five minutes.

Thus we see that the beginning of the neutrino radiation from the hot central regions of contracting stars gives us the complete explanation of the causes of stellar collapses.

It must be stated, however, that although the rate of energy losses through neutrino emission can be estimated comparatively easily, the study of the collapse process itself presents many mathematical difficulties, so that only the qualitative explanation of the events can be given at present.

It must be imagined that, as the result of the deficiency of gas pressure in stellar interiors, the masses that form its giant outside body begin to fall toward the center driven by the forces of gravity. Since, however, every star is usually in a state of more or less rapid rotation, the process of the collapse proceeds asymmetrically, and the polar masses (i.e., those located near the axis of rotation) fall in first pushing the equatorial masses outward (Figure 126).

This brings out the material previously hidden deep in the stellar interior, and heated up to the temperatures of several thousand million degrees, a temperature which accounts for the sudden increase of stellar luminosity. As the process goes on the collapsing material of the old star condenses in the center into a dense white dwarf star, whereas the expelled masses gradually cool down and continue to expand forming the sort of nebulosity observed in the Crab Nebula.

3. PRIMORDIAL CHAOS AND THE EXPANDING UNIVERSE

Thinking of the universe as a whole, we are at once confronted by vital questions concerning its possible evolution in time. Must we assume that it always was, and will always remain, in approximately the same state as we observe it now? Or does the universe continuously change, passing through different evolutionary stages?

Examining this question on the basis of empirical facts collected from widely different branches of science, we come to a quite definite answer. *Yes, our universe is gradually changing*; its state in the long-forgotten past, its state in the present, and that which it will become in the distant future are three very different states of being. The numerous facts collected by various sciences indicate furthermore that our universe had a certain *beginning*, from which it developed into its present state through the process of gradual evolution. As we have seen above, the age of our planetary system can be estimated as a few billion years, a figure that emerges stubbornly from many independent attacks on the problem from several directions. The formation of the moon, which was apparently torn away from the body of the Earth by vigorous gravitational forces emanating from the sun, also must have taken place a few billion years ago.

The study of the evolution of individual stars (see the previous section) indicates that most of the stars that we see now in the sky are also *several billion years* old. The study of the motion of stars in general, and in particular the relative motion of double and triple stellar systems, as well as that of the more complicated stellar groups known as *galactic clusters*, leads astronomers to the conclusion that such configurations could not have existed for a longer time than this.

Quite independent evidence is supplied by considerations of the relative abundance of various chemical elements, and in particular of the amounts of radioactive elements such as thorium and uranium that are known to be decaying gradually. If, in spite of their progressive decay, these elements are still present in the universe, we must assume either that they are continuously produced from other lighter nuclei even at the present time, or that they are the last remnants of a stock pile formed by nature in some distant past.

Our present knowledge of nuclear transformation processes forces us to abandon the first possibility, since even in the interior of the hottest stars the temperature never rises to the tremendous heights necessary for "cooking" the heavy radioactive nuclei. In fact, as we have seen in the preceding section the temperatures in the interior of stars are measured in tens of millions of degrees, whereas several billion degrees are needed to "cook" radioactive nuclei from the nuclei of lighter elements.

Accordingly, we must assume that the nuclei of heavy elements were formed in some past epoch of the evolution of the universe, and that *at that particular epoch all matter was subjected to some terrifically high temperatures and correspondingly high pressures.*

We can also arrive at an estimate of the approximate date of this "purgatory" stage of the universe. We know that thorium and uranium-238, which have mean life spans of 18 and $4\frac{1}{2}$ billion years respectively, have not decayed materially since they were formed, for they are at present about as abundant as some other stable heavy elements. On the other hand, uranium-235, with a mean life span of only about a half billion years, is 140 times less abundant than uranium-238. The present large abundance of uranium-238 and thorium indicates that the formation of elements could not have taken place more than a few billion years ago, and the small amount of uranium-235 makes a still closer estimate possible. In fact, if the amount of this element was halved every 500 million years, it must have taken about seven such periods, that is, $3\frac{1}{2}$ billion years to cut it down to one 140th (since $\frac{1}{2} \times \frac{1}{2} \times \frac{1}{2} \times \frac{1}{2} \times \frac{1}{2} \times \frac{1}{2} \times \frac{1}{2} = \frac{1}{128}$).

This estimate of the age of chemical elements, obtained exclusively from the data of nuclear physics, is in beautiful agreement with the estimated age of planets, stars, and stellar groups obtained from purely astronomical data!

But what was the state of the universe during that early age, several billion years ago when everything seems to have been formed? And what are the changes which have taken place in the meantime to bring the universe to its present state?

The most complete answer to the above questions can be obtained from the study of the phenomenon of "universal expansion." We have seen in the previous chapter that the vast space of the universe is filled by a large number of giant stellar systems or galaxies, and that our sun is only one of many billions of stars in one of such galaxies, known as the Milky Way. We have also seen that these galaxies are more or less uniformly scattered through space as far as the eye (helped, of course, by the 200-inch telescope) can see.

Studying the spectra of the light coming from these distant galaxies, Mt. Wilson's astronomer E. Hubble noticed that the spectral lines are shifted slightly towards the red end of the spectrum, and that this so-called "red shift" is stronger in the more distant galaxies. In fact, it was found that the "red shift" observed in different galaxies is directly proportional to their distance from us.

The most natural way to explain this phenomenon is to assume that *all galaxies recede from us with a speed that increases with their distance from us.* This explanation is based on the so-called "Doppler effect," which makes the light coming from a source that is approaching us change its color toward the violet end of the spectrum, and light from a receding source change toward the red. Of course, to obtain a noticeable shift the relative ve-

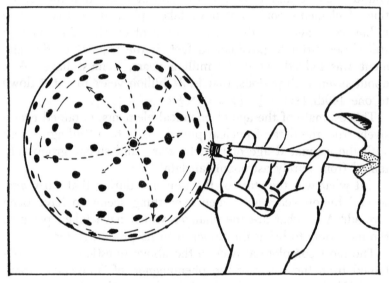

FIGURE 127

The dots run away from one another when the rubber balloon is expanding.

locity of the source in relation to the position of the observer must be rather large. When Prof. R. W. Wood was arrested for going through a red traffic signal in Baltimore and told the judge that, because of this phenomenon, the light he saw looked green to him, since he was approaching it in the car, the professor was simply pulling the judge's leg. Had the judge known more about physics, he would have asked Professor Wood to calculate the speed with which he must have been driving in order to see green in a red light, and then would have fined him for speeding!

Returning to the problem of the "red shift" observed in galaxies, we come to what is at first sight a rather awkward conclusion. It looks as though all of the galaxies in the universe were running away from our Milky Way as if it were a galactic Monster of Frankenstein! What then are the horrible properties of our own stellar system, and why does it seem to be so unpopular among all other galaxies? If you think a little about this question, you will easily come to the conclusion that there is nothing particularly wrong with our Milky Way, and that, in fact, other galaxies do not run away from it exclusively but rather that all run away from one another. Imagine a rubber balloon with a polka-dot pattern painted on its surface (Figure 127). If you will begin to inflate it, gradually stretching its surface to a large and larger size, the distances between individual dots will continually increase so that an insect sitting on any one of the dots would receive the impression that all other other dots are "running away" from it. Moreover the recession velocities of different dots on the expanding balloon will be directly proportional to their distance from the insect's observation point.

This example must make it quite clear that the recession of galaxies observed by Hubble has nothing to do with the particular properties or position of our own galaxy, but must be interpreted simply as due to *the general uniform expansion of the system of galaxies scattered through the space of the universe.*

From the observed velocity of the expansion and the present distances between the neighboring galaxies one easily calculates that *this expansion must have started more than five billion years ago.*[6]

Before that time the separate stellar clouds that we now call galaxies were forming the parts of a uniform distribution of stars

---

[6] According to Hubble's original data, the mean distance between two neighboring galaxies is about 1.7 million light-years (or $1 \cdot 6 \cdot 10^{19}$ km), whereas their mutual recession velocity is about 300 km per second. Assuming a uniform expansion rate we obtain for the expansion time $\dfrac{1 \cdot 6 \cdot 10^{19}}{300} = 5 \cdot 10^{16}$ sec $= 1 \cdot 8 \cdot 10^{9}$ yr. More recent information leads, however, to an estimate of somewhat longer time periods.

through the entire space of the universe, and at a still earlier date, the stars themselves were squeezed together filling the universe with continuously distributed hot gas. Going still farther back in time we find that this gas was denser and hotter, and this was apparently an epoch when different chemical (and particularly the radioactive) elements were formed. One more step back in time and we find that the matter of the universe was squeezed into the superdense superheated nuclear fluid discussed in Chapter VII.

Now we can assemble these observations and see the events which marked the evolutionary development of the universe in their correct order.

The story begins with the embryonic stage of the universe when all the matter that we can see now scattered through space to the limits of vision of the Mt. Wilson telescope (i.e., within a radius of 500,000,000 light-years) was squeezed into a sphere with a radius of only about eight sun radii.[7] However, this extra dense state did not last very long, since rapid expansion must have brought the density of the universe down to a million times the density of water within the first two seconds, and to that of water density within a few hours. Approximately at this time the previously continuous gas must have been broken into separate gaseous spheres that now constitute individual stars. These stars being pulled apart by the progressing expansion broke up later into separate stellar clouds, which we call galaxies and which are still receding from one another into the unknown depths of the universe.

Now we can ask ourselves what kind of forces are responsible for the expansion of the universe, and whether this expansion will ever stop or even become contraction. Is there any possibility that the expanding masses of the universe will turn back on us

---

[7] Since the density of nuclear fluid is $10^{14} \frac{gm}{cm^3}$, and the present mean density of matter in space is $10^{-30} \frac{gm}{cm^3}$, the linear contraction was $\sqrt[3]{\frac{10^{14}}{10^{-30}}} = 5 \cdot 10^{14}$. Thus the present distances of $5 \cdot 10^8$ light-years were at that time only $\frac{5 \cdot 10^8}{5 \cdot 10^{14}} = 10^{-6}$ light-years $= 10,000.000$ km.

and squeeze our stellar system, the Milky Way, the sun, the Earth, and the humanity on earth into a pulp with nuclear density?

According to conclusions based upon the best available information, this will never happen. Long ago, in the early stages of its evolution, the expanding universe broke all of the ties that might have held it together and is now expanding into infinity obedient to the simple law of inertia. The ties we have just mentioned were formed by gravity forces which tended to prevent the masses of the universe from drawing apart.

FIGURE 128

To construct a simple explanatory example, let us suppose that we try to shoot a rocket from the surface of the Earth into interplanetary space. We know that no existing rockets, not even the famous V2, have enough propulsive power to escape into the free space, that they are always stopped in their ascension by the forces of gravity and pulled back to Earth. However, if we were able to power a rocket so that it would leave the earth with an initial speed in excess of 11 km per second (which seems a goal possible of achievement in the development of atomic-jet-propelled rockets), it will be able to push beyond the pull of earth's gravity and to escape int） the free space, where it will continue to move without hindrance. The velocity of 11 km per second is usually known as the "escape velocity" from the gravity of the Earth.

Imagine now an artillery shell that has exploded in midair, sending its fragments in all directions (Figure 128a). The fragments thrown out by the force of the explosion fly apart against the gravitational forces that tend to pull them back toward the common center. It goes without saying that in the case of shell fragments, these forces of mutual gravitational attraction are negligible, that is, they are so weak as not to influence at all the motion of the fragments through space. If, however, these forces were stronger, they would be able to stop the fragments in their flight, and to make them fall back, to their common center of gravity (Figure 128*b*). The question as to whether the fragments will come back or fly into infinity is decided by the relative values of their kinetic energy of motion, and the potential energy of gravity forces between them.

Substitute for the shell fragments the separate galaxies, and you will have a picture of the expanding universe, as described on the previous pages. Here, however, because of very large masses of individual fragment galaxies, the potential energy of gravitational forces becomes quite important as compared with their kinetic energy,[8] so that the future of the expansion can be decided only by a careful study of the two quantities involved.

According to the best available information concerning the galactic masses, it seems that at present the kinetic energy of receding galaxies is several times greater than their mutual potential gravitational energy, from which it would follow that *our universe is expanding into infinity without any chance of ever being pulled more closely together again by the forces of gravity.* It must be remembered, however, that most of the numerical data pertaining to the universe as a whole are not very exact, and it is possible that future studies will reverse this conclusion. But even if the expanding universe does suddenly stop in its tracks, and turn back in a movement of compression, it will be billions of years before that terrible day envisioned by the Negro Spiritual, "when the stars begin to fall," and we are crushed under the weight of collapsing galaxies!

What was this high explosive material that sent the fragments of the universe flying apart at such a terrific speed? The answer may be somewhat disappointing: there probably was no explosion in the ordinary sense of the word. The universe is now

[8] Whereas kinetic energy of moving particles is proportional to their mass, their mutual potential energy increases as the square of their masses.

expanding because in some previous period of its history (of which, of course, no record has been left), it contracted from infinity into a very dense state and then rebounded, as it were, propelled by the strong elastic forces inherent in compressed matter. If you were to enter a game room just in time to see a pingpong ball rising from the floor high into the air, you would conclude (without really thinking about it) that in the instant before you entered the room the ball had fallen to the floor from a comparable height, and was jumping up again because of its elasticity.

We can now send our imagination flying beyond any limits, and ask ourselves whether during the precompressive stages of the universe everything that is now happening was happening in reverse order.

Were you reading this book from the last page to the first some eight or ten billion years ago? And did the people of that time produce fried chickens from their mouth, put life into them in the kitchen, and send them to the farm where they grew from adulthood to babyhood, finally crawled into eggshells, and after some weeks became fresh eggs? Interesting as they are, such questions cannot be answered from the purely scientific point of view, since the maximum compression of the universe, which squeezed all matter into a uniform nuclear fluid, must have completely obliterated all the records of the earlier compressive stages.

Index

A CATALOG OF SELECTED
DOVER BOOKS
IN ALL FIELDS OF INTEREST

A CATALOG OF SELECTED DOVER
BOOKS IN ALL FIELDS OF INTEREST

CONCERNING THE SPIRITUAL IN ART, Wassily Kandinsky. Pioneering work by father of abstract art. Thoughts on color theory, nature of art. Analysis of earlier masters. 12 illustrations. 80pp. of text. 5⅜ x 8½. 0-486-23411-8

CELTIC ART: The Methods of Construction, George Bain. Simple geometric techniques for making Celtic interlacements, spirals, Kells-type initials, animals, humans, etc. Over 500 illustrations. 160pp. 9 x 12. (Available in U.S. only.) 0-486-22923-8

AN ATLAS OF ANATOMY FOR ARTISTS, Fritz Schider. Most thorough reference work on art anatomy in the world. Hundreds of illustrations, including selections from works by Vesalius, Leonardo, Goya, Ingres, Michelangelo, others. 593 illustrations. 192pp. 7⅛ x 10¼. 0-486-20241-0

CELTIC HAND STROKE-BY-STROKE (Irish Half-Uncial from "The Book of Kells"): An Arthur Baker Calligraphy Manual, Arthur Baker. Complete guide to creating each letter of the alphabet in distinctive Celtic manner. Covers hand position, strokes, pens, inks, paper, more. Illustrated. 48pp. 8¼ x 11. 0-486-24336-2

EASY ORIGAMI, John Montroll. Charming collection of 32 projects (hat, cup, pelican, piano, swan, many more) specially designed for the novice origami hobbyist. Clearly illustrated easy-to-follow instructions insure that even beginning papercrafters will achieve successful results. 48pp. 8¼ x 11. 0-486-27298-2

BLOOMINGDALE'S ILLUSTRATED 1886 CATALOG: Fashions, Dry Goods and Housewares, Bloomingdale Brothers. Famed merchants' extremely rare catalog depicting about 1,700 products: clothing, housewares, firearms, dry goods, jewelry, more. Invaluable for dating, identifying vintage items. Also, copyright-free graphics for artists, designers. Co-published with Henry Ford Museum & Greenfield Village. 160pp. 8¼ x 11. 0-486-25780-0

THE ART OF WORLDLY WISDOM, Baltasar Gracian. "Think with the few and speak with the many," "Friends are a second existence," and "Be able to forget" are among this 1637 volume's 300 pithy maxims. A perfect source of mental and spiritual refreshment, it can be opened at random and appreciated either in brief or at length. 128pp. 5⅜ x 8½. 0-486-44034-6

JOHNSON'S DICTIONARY: A Modern Selection, Samuel Johnson (E. L. McAdam and George Milne, eds.). This modern version reduces the original 1755 edition's 2,300 pages of definitions and literary examples to a more manageable length, retaining the verbal pleasure and historical curiosity of the original. 480pp. 5³⁄₁₆ x 8¼. 0-486-44089-3

ADVENTURES OF HUCKLEBERRY FINN, Mark Twain, Illustrated by E. W. Kemble. A work of eternal richness and complexity, a source of ongoing critical debate, and a literary landmark, Twain's 1885 masterpiece about a barefoot boy's journey of self-discovery has enthralled readers around the world. This handsome clothbound reproduction of the first edition features all 174 of the original black-and-white illustrations. 368pp. 5⅜ x 8½. 0-486-44322-1

STICKLEY CRAFTSMAN FURNITURE CATALOGS, Gustav Stickley and L. & J. G. Stickley. Beautiful, functional furniture in two authentic catalogs from 1910. 594 illustrations, including 277 photos, show settles, rockers, armchairs, reclining chairs, bookcases, desks, tables. 183pp. 6½ x 9¼. 0-486-23838-5

AMERICAN LOCOMOTIVES IN HISTORIC PHOTOGRAPHS: 1858 to 1949, Ron Ziel (ed.). A rare collection of 126 meticulously detailed official photographs, called "builder portraits," of American locomotives that majestically chronicle the rise of steam locomotive power in America. Introduction. Detailed captions. xi+ 129pp. 9 x 12. 0-486-27393-8

AMERICA'S LIGHTHOUSES: An Illustrated History, Francis Ross Holland, Jr. Delightfully written, profusely illustrated fact-filled survey of over 200 American lighthouses since 1716. History, anecdotes, technological advances, more. 240pp. 8 x 10¾.
0-486-25576-X

TOWARDS A NEW ARCHITECTURE, Le Corbusier. Pioneering manifesto by founder of "International School." Technical and aesthetic theories, views of industry, economics, relation of form to function, "mass-production split" and much more. Profusely illustrated. 320pp. 6⅛ x 9¼. (Available in U.S. only.) 0-486-25023-7

HOW THE OTHER HALF LIVES, Jacob Riis. Famous journalistic record, exposing poverty and degradation of New York slums around 1900, by major social reformer. 100 striking and influential photographs. 233pp. 10 x 7⅞. 0-486-22012-5

FRUIT KEY AND TWIG KEY TO TREES AND SHRUBS, William M. Harlow. One of the handiest and most widely used identification aids. Fruit key covers 120 deciduous and evergreen species; twig key 160 deciduous species. Easily used. Over 300 photographs. 126pp. 5⅜ x 8½. 0-486-20511-8

COMMON BIRD SONGS, Dr. Donald J. Borror. Songs of 60 most common U.S. birds: robins, sparrows, cardinals, bluejays, finches, more—arranged in order of increasing complexity. Up to 9 variations of songs of each species.
Cassette and manual 0-486-99911-4

ORCHIDS AS HOUSE PLANTS, Rebecca Tyson Northen. Grow cattleyas and many other kinds of orchids—in a window, in a case, or under artificial light. 63 illustrations. 148pp. 5⅜ x 8½. 0-486-23261-1

MONSTER MAZES, Dave Phillips. Masterful mazes at four levels of difficulty. Avoid deadly perils and evil creatures to find magical treasures. Solutions for all 32 exciting illustrated puzzles. 48pp. 8¼ x 11. 0-486-26005-4

MOZART'S DON GIOVANNI (DOVER OPERA LIBRETTO SERIES), Wolfgang Amadeus Mozart. Introduced and translated by Ellen H. Bleiler. Standard Italian libretto, with complete English translation. Convenient and thoroughly portable—an ideal companion for reading along with a recording or the performance itself. Introduction. List of characters. Plot summary. 121pp. 5¼ x 8½. 0-486-24944-1

FRANK LLOYD WRIGHT'S DANA HOUSE, Donald Hoffmann. Pictorial essay of residential masterpiece with over 160 interior and exterior photos, plans, elevations, sketches and studies. 128pp. 9¼ x 10¾. 0-486-29120-0

THE CLARINET AND CLARINET PLAYING, David Pino. Lively, comprehensive work features suggestions about technique, musicianship, and musical interpretation, as well as guidelines for teaching, making your own reeds, and preparing for public performance. Includes an intriguing look at clarinet history. "A godsend," *The Clarinet,* Journal of the International Clarinet Society. Appendixes. 7 illus. 320pp. $5\frac{3}{8} \times 8\frac{1}{2}$. 0-486-40270-3

HOLLYWOOD GLAMOR PORTRAITS, John Kobal (ed.). 145 photos from 1926-49. Harlow, Gable, Bogart, Bacall; 94 stars in all. Full background on photographers, technical aspects. 160pp. $8\frac{3}{8} \times 11\frac{1}{4}$. 0-486-23352-9

THE RAVEN AND OTHER FAVORITE POEMS, Edgar Allan Poe. Over 40 of the author's most memorable poems: "The Bells," "Ulalume," "Israfel," "To Helen," "The Conqueror Worm," "Eldorado," "Annabel Lee," many more. Alphabetic lists of titles and first lines. 64pp. $5\frac{3}{16} \times 8\frac{1}{4}$. 0-486-26685-0

PERSONAL MEMOIRS OF U. S. GRANT, Ulysses Simpson Grant. Intelligent, deeply moving firsthand account of Civil War campaigns, considered by many the finest military memoirs ever written. Includes letters, historic photographs, maps and more. 528pp. $6\frac{1}{8} \times 9\frac{1}{4}$. 0-486-28587-1

ANCIENT EGYPTIAN MATERIALS AND INDUSTRIES, A. Lucas and J. Harris. Fascinating, comprehensive, thoroughly documented text describes this ancient civilization's vast resources and the processes that incorporated them in daily life, including the use of animal products, building materials, cosmetics, perfumes and incense, fibers, glazed ware, glass and its manufacture, materials used in the mummification process, and much more. 544pp. $6^1/_8 \times 9^1/_4$. (Available in U.S. only.) 0-486-40446-3

RUSSIAN STORIES/RUSSKIE RASSKAZY: A Dual-Language Book, edited by Gleb Struve. Twelve tales by such masters as Chekhov, Tolstoy, Dostoevsky, Pushkin, others. Excellent word-for-word English translations on facing pages, plus teaching and study aids, Russian/English vocabulary, biographical/critical introductions, more. 416pp. $5\frac{3}{8} \times 8\frac{1}{2}$. 0-486-26244-8

PHILADELPHIA THEN AND NOW: 60 Sites Photographed in the Past and Present, Kenneth Finkel and Susan Oyama. Rare photographs of City Hall, Logan Square, Independence Hall, Betsy Ross House, other landmarks juxtaposed with contemporary views. Captures changing face of historic city. Introduction. Captions. 128pp. $8\frac{1}{4} \times 11$. 0-486-25790-8

NORTH AMERICAN INDIAN LIFE: Customs and Traditions of 23 Tribes, Elsie Clews Parsons (ed.). 27 fictionalized essays by noted anthropologists examine religion, customs, government, additional facets of life among the Winnebago, Crow, Zuni, Eskimo, other tribes. 480pp. $6\frac{1}{8} \times 9\frac{1}{4}$. 0-486-27377-6

TECHNICAL MANUAL AND DICTIONARY OF CLASSICAL BALLET, Gail Grant. Defines, explains, comments on steps, movements, poses and concepts. 15-page pictorial section. Basic book for student, viewer. 127pp. $5\frac{3}{8} \times 8\frac{1}{2}$. 0-486-21843-0

THE MALE AND FEMALE FIGURE IN MOTION: 60 Classic Photographic Sequences, Eadweard Muybridge. 60 true-action photographs of men and women walking, running, climbing, bending, turning, etc., reproduced from rare 19th-century masterpiece. vi + 121pp. 9×12. 0-486-24745-7

ANIMALS: 1,419 Copyright-Free Illustrations of Mammals, Birds, Fish, Insects, etc., Jim Harter (ed.). Clear wood engravings present, in extremely lifelike poses, over 1,000 species of animals. One of the most extensive pictorial sourcebooks of its kind. Captions. Index. 284pp. 9 x 12. 0-486-23766-4

1001 QUESTIONS ANSWERED ABOUT THE SEASHORE, N. J. Berrill and Jacquelyn Berrill. Queries answered about dolphins, sea snails, sponges, starfish, fishes, shore birds, many others. Covers appearance, breeding, growth, feeding, much more. 305pp. 5¼ x 8¼. 0-486-23366-9

ATTRACTING BIRDS TO YOUR YARD, William J. Weber. Easy-to-follow guide offers advice on how to attract the greatest diversity of birds: birdhouses, feeders, water and waterers, much more. 96pp. 5³⁄₁₆ x 8¼. 0-486-28927-3

MEDICINAL AND OTHER USES OF NORTH AMERICAN PLANTS: A Historical Survey with Special Reference to the Eastern Indian Tribes, Charlotte Erichsen-Brown. Chronological historical citations document 500 years of usage of plants, trees, shrubs native to eastern Canada, northeastern U.S. Also complete identifying information. 343 illustrations. 544pp. 6½ x 9¼. 0-486-25951-X

STORYBOOK MAZES, Dave Phillips. 23 stories and mazes on two-page spreads: Wizard of Oz, Treasure Island, Robin Hood, etc. Solutions. 64pp. 8¼ x 11. 0-486-23628-5

AMERICAN NEGRO SONGS: 230 Folk Songs and Spirituals, Religious and Secular, John W. Work. This authoritative study traces the African influences of songs sung and played by black Americans at work, in church, and as entertainment. The author discusses the lyric significance of such songs as "Swing Low, Sweet Chariot," "John Henry," and others and offers the words and music for 230 songs. Bibliography. Index of Song Titles. 272pp. 6½ x 9¼. 0-486-40271-1

MOVIE-STAR PORTRAITS OF THE FORTIES, John Kobal (ed.). 163 glamor, studio photos of 106 stars of the 1940s: Rita Hayworth, Ava Gardner, Marlon Brando, Clark Gable, many more. 176pp. 8⅜ x 11¼. 0-486-23546-7

YEKL and THE IMPORTED BRIDEGROOM AND OTHER STORIES OF YIDDISH NEW YORK, Abraham Cahan. Film Hester Street based on *Yekl* (1896). Novel, other stories among first about Jewish immigrants on N.Y.'s East Side. 240pp. 5⅜ x 8½. 0-486-22427-9

SELECTED POEMS, Walt Whitman. Generous sampling from *Leaves of Grass*. Twenty-four poems include "I Hear America Singing," "Song of the Open Road," "I Sing the Body Electric," "When Lilacs Last in the Dooryard Bloom'd," "O Captain! My Captain!"–all reprinted from an authoritative edition. Lists of titles and first lines. 128pp. 5³⁄₁₆ x 8¼. 0-486-26878-0

SONGS OF EXPERIENCE: Facsimile Reproduction with 26 Plates in Full Color, William Blake. 26 full-color plates from a rare 1826 edition. Includes "The Tyger," "London," "Holy Thursday," and other poems. Printed text of poems. 48pp. 5¼ x 7. 0-486-24636-1

THE BEST TALES OF HOFFMANN, E. T. A. Hoffmann. 10 of Hoffmann's most important stories: "Nutcracker and the King of Mice," "The Golden Flowerpot," etc. 458pp. 5⅜ x 8½. 0-486-21793-0

THE BOOK OF TEA, Kakuzo Okakura. Minor classic of the Orient: entertaining, charming explanation, interpretation of traditional Japanese culture in terms of tea ceremony. 94pp. 5⅜ x 8½. 0-486-20070-1

FRENCH STORIES/CONTES FRANÇAIS: A Dual-Language Book, Wallace Fowlie. Ten stories by French masters, Voltaire to Camus: "Micromegas" by Voltaire; "The Atheist's Mass" by Balzac; "Minuet" by de Maupassant; "The Guest" by Camus, six more. Excellent English translations on facing pages. Also French-English vocabulary list, exercises, more. 352pp. 5⅜ x 8½.   0-486-26443-2

CHICAGO AT THE TURN OF THE CENTURY IN PHOTOGRAPHS: 122 Historic Views from the Collections of the Chicago Historical Society, Larry A. Viskochil. Rare large-format prints offer detailed views of City Hall, State Street, the Loop, Hull House, Union Station, many other landmarks, circa 1904-1913. Introduction. Captions. Maps. 144pp. 9⅜ x 12¼.   0-486-24656-6

OLD BROOKLYN IN EARLY PHOTOGRAPHS, 1865-1929, William Lee Younger. Luna Park, Gravesend race track, construction of Grand Army Plaza, moving of Hotel Brighton, etc. 157 previously unpublished photographs. 165pp. 8⅞ x 11¾.
0-486-23587-4

THE MYTHS OF THE NORTH AMERICAN INDIANS, Lewis Spence. Rich anthology of the myths and legends of the Algonquins, Iroquois, Pawnees and Sioux, prefaced by an extensive historical and ethnological commentary. 36 illustrations. 480pp. 5⅜ x 8½.   0-486-25967-6

AN ENCYCLOPEDIA OF BATTLES: Accounts of Over 1,560 Battles from 1479 B.C. to the Present, David Eggenberger. Essential details of every major battle in recorded history from the first battle of Megiddo in 1479 B.C. to Grenada in 1984. List of Battle Maps. New Appendix covering the years 1967-1984. Index. 99 illustrations. 544pp. 6½ x 9¼.   0-486-24913-1

SAILING ALONE AROUND THE WORLD, Captain Joshua Slocum. First man to sail around the world, alone, in small boat. One of great feats of seamanship told in delightful manner. 67 illustrations. 294pp. 5⅜ x 8½.   0-486-20326-3

ANARCHISM AND OTHER ESSAYS, Emma Goldman. Powerful, penetrating, prophetic essays on direct action, role of minorities, prison reform, puritan hypocrisy, violence, etc. 271pp. 5⅜ x 8½.   0-486-22484-8

MYTHS OF THE HINDUS AND BUDDHISTS, Ananda K. Coomaraswamy and Sister Nivedita. Great stories of the epics; deeds of Krishna, Shiva, taken from puranas, Vedas, folk tales; etc. 32 illustrations. 400pp. 5⅜ x 8½.   0-486-21759-0

MY BONDAGE AND MY FREEDOM, Frederick Douglass. Born a slave, Douglass became outspoken force in antislavery movement. The best of Douglass' autobiographies. Graphic description of slave life. 464pp. 5⅜ x 8½.   0-486-22457-0

FOLLOWING THE EQUATOR: A Journey Around the World, Mark Twain. Fascinating humorous account of 1897 voyage to Hawaii, Australia, India, New Zealand, etc. Ironic, bemused reports on peoples, customs, climate, flora and fauna, politics, much more. 197 illustrations. 720pp. 5⅜ x 8½.   0-486-26113-1

THE PEOPLE CALLED SHAKERS, Edward D. Andrews. Definitive study of Shakers: origins, beliefs, practices, dances, social organization, furniture and crafts, etc. 33 illustrations. 351pp. 5⅜ x 8½.   0-486-21081-2

THE MYTHS OF GREECE AND ROME, H. A. Guerber. A classic of mythology, generously illustrated, long prized for its simple, graphic, accurate retelling of the principal myths of Greece and Rome, and for its commentary on their origins and significance. With 64 illustrations by Michelangelo, Raphael, Titian, Rubens, Canova, Bernini and others. 480pp. 5⅜ x 8½.   0-486-27584-1

PSYCHOLOGY OF MUSIC, Carl E. Seashore. Classic work discusses music as a medium from psychological viewpoint. Clear treatment of physical acoustics, auditory apparatus, sound perception, development of musical skills, nature of musical feeling, host of other topics. 88 figures. 408pp. 5⅜ x 8½. 0-486-21851-1

LIFE IN ANCIENT EGYPT, Adolf Erman. Fullest, most thorough, detailed older account with much not in more recent books, domestic life, religion, magic, medicine, commerce, much more. Many illustrations reproduce tomb paintings, carvings, hieroglyphs, etc. 597pp. 5⅜ x 8½. 0-486-22632-8

SUNDIALS, Their Theory and Construction, Albert Waugh. Far and away the best, most thorough coverage of ideas, mathematics concerned, types, construction, adjusting anywhere. Simple, nontechnical treatment allows even children to build several of these dials. Over 100 illustrations. 230pp. 5⅜ x 8½. 0-486-22947-5

THEORETICAL HYDRODYNAMICS, L. M. Milne-Thomson. Classic exposition of the mathematical theory of fluid motion, applicable to both hydrodynamics and aerodynamics. Over 600 exercises. 768pp. 6⅛ x 9¼. 0-486-68970-0

OLD-TIME VIGNETTES IN FULL COLOR, Carol Belanger Grafton (ed.). Over 390 charming, often sentimental illustrations, selected from archives of Victorian graphics–pretty women posing, children playing, food, flowers, kittens and puppies, smiling cherubs, birds and butterflies, much more. All copyright-free. 48pp. 9¼ x 12¼. 0-486-27269-9

PERSPECTIVE FOR ARTISTS, Rex Vicat Cole. Depth, perspective of sky and sea, shadows, much more, not usually covered. 391 diagrams, 81 reproductions of drawings and paintings. 279pp. 5⅜ x 8½. 0-486-22487-2

DRAWING THE LIVING FIGURE, Joseph Sheppard. Innovative approach to artistic anatomy focuses on specifics of surface anatomy, rather than muscles and bones. Over 170 drawings of live models in front, back and side views, and in widely varying poses. Accompanying diagrams. 177 illustrations. Introduction. Index. 144pp. 8⅜ x11¼. 0-486-26723-7

GOTHIC AND OLD ENGLISH ALPHABETS: 100 Complete Fonts, Dan X. Solo. Add power, elegance to posters, signs, other graphics with 100 stunning copyright-free alphabets: Blackstone, Dolbey, Germania, 97 more–including many lower-case, numerals, punctuation marks. 104pp. 8⅛ x 11. 0-486-24695-7

THE BOOK OF WOOD CARVING, Charles Marshall Sayers. Finest book for beginners discusses fundamentals and offers 34 designs. "Absolutely first rate . . . well thought out and well executed."–E. J. Tangerman. 118pp. 7¾ x 10⅝. 0-486-23654-4

ILLUSTRATED CATALOG OF CIVIL WAR MILITARY GOODS: Union Army Weapons, Insignia, Uniform Accessories, and Other Equipment, Schuyler, Hartley, and Graham. Rare, profusely illustrated 1846 catalog includes Union Army uniform and dress regulations, arms and ammunition, coats, insignia, flags, swords, rifles, etc. 226 illustrations. 160pp. 9 x 12. 0-486-24939-5

WOMEN'S FASHIONS OF THE EARLY 1900s: An Unabridged Republication of "New York Fashions, 1909," National Cloak & Suit Co. Rare catalog of mail-order fashions documents women's and children's clothing styles shortly after the turn of the century. Captions offer full descriptions, prices. Invaluable resource for fashion, costume historians. Approximately 725 illustrations. 128pp. 8⅜ x 11¼. 0-486-27276-1

HOW TO DO BEADWORK, Mary White. Fundamental book on craft from simple projects to five-bead chains and woven works. 106 illustrations. 142pp. 5⅜ x 8. 0-486-20697-1

THE 1912 AND 1915 GUSTAV STICKLEY FURNITURE CATALOGS, Gustav Stickley. With over 200 detailed illustrations and descriptions, these two catalogs are essential reading and reference materials and identification guides for Stickley furniture. Captions cite materials, dimensions and prices. 112pp. 6½ x 9¼. 0-486-26676-1

EARLY AMERICAN LOCOMOTIVES, John H. White, Jr. Finest locomotive engravings from early 19th century: historical (1804–74), main-line (after 1870), special, foreign, etc. 147 plates. 142pp. 11⅜ x 8¼. 0-486-22772-3

LITTLE BOOK OF EARLY AMERICAN CRAFTS AND TRADES, Peter Stockham (ed.). 1807 children's book explains crafts and trades: baker, hatter, cooper, potter, and many others. 23 copperplate illustrations. 140pp. 4⅝ x 6. 0-486-23336-7

VICTORIAN FASHIONS AND COSTUMES FROM HARPER'S BAZAR, 1867–1898, Stella Blum (ed.). Day costumes, evening wear, sports clothes, shoes, hats, other accessories in over 1,000 detailed engravings. 320pp. 9⅜ x 12¼. 0-486-22990-4

THE LONG ISLAND RAIL ROAD IN EARLY PHOTOGRAPHS, Ron Ziel. Over 220 rare photos, informative text document origin (1844) and development of rail service on Long Island. Vintage views of early trains, locomotives, stations, passengers, crews, much more. Captions. 8⅞ x 11¾. 0-486-26301-0

VOYAGE OF THE LIBERDADE, Joshua Slocum. Great 19th-century mariner's thrilling, first-hand account of the wreck of his ship off South America, the 35-foot boat he built from the wreckage, and its remarkable voyage home. 128pp. 5⅜ x 8½. 0-486-40022-0

TEN BOOKS ON ARCHITECTURE, Vitruvius. The most important book ever written on architecture. Early Roman aesthetics, technology, classical orders, site selection, all other aspects. Morgan translation. 331pp. 5⅜ x 8½. 0-486-20645-9

THE HUMAN FIGURE IN MOTION, Eadweard Muybridge. More than 4,500 stopped-action photos, in action series, showing undraped men, women, children jumping, lying down, throwing, sitting, wrestling, carrying, etc. 390pp. 7⅞ x 10⅝. 0-486-20204-6 Clothbd.

TREES OF THE EASTERN AND CENTRAL UNITED STATES AND CANADA, William M. Harlow. Best one-volume guide to 140 trees. Full descriptions, woodlore, range, etc. Over 600 illustrations. Handy size. 288pp. 4½ x 6⅜. 0-486-20395-6

GROWING AND USING HERBS AND SPICES, Milo Miloradovich. Versatile handbook provides all the information needed for cultivation and use of all the herbs and spices available in North America. 4 illustrations. Index. Glossary. 236pp. 5⅜ x 8½. 0-486-25058-X

BIG BOOK OF MAZES AND LABYRINTHS, Walter Shepherd. 50 mazes and labyrinths in all–classical, solid, ripple, and more–in one great volume. Perfect inexpensive puzzler for clever youngsters. Full solutions. 112pp. 8⅛ x 11. 0-486-22951-3

PIANO TUNING, J. Cree Fischer. Clearest, best book for beginner, amateur. Simple repairs, raising dropped notes, tuning by easy method of flattened fifths. No previous skills needed. 4 illustrations. 201pp. 5⅜ x 8½. 0-486-23267-0

HINTS TO SINGERS, Lillian Nordica. Selecting the right teacher, developing confidence, overcoming stage fright, and many other important skills receive thoughtful discussion in this indispensible guide, written by a world-famous diva of four decades' experience. 96pp. 5⅜ x 8½. 0-486-40094-8

THE COMPLETE NONSENSE OF EDWARD LEAR, Edward Lear. All nonsense limericks, zany alphabets, Owl and Pussycat, songs, nonsense botany, etc., illustrated by Lear. Total of 320pp. 5⅜ x 8½. (Available in U.S. only.) 0-486-20167-8

VICTORIAN PARLOUR POETRY: An Annotated Anthology, Michael R. Turner. 117 gems by Longfellow, Tennyson, Browning, many lesser-known poets. "The Village Blacksmith," "Curfew Must Not Ring Tonight," "Only a Baby Small," dozens more, often difficult to find elsewhere. Index of poets, titles, first lines. xxiii + 325pp. 5⅝ x 8¼. 0-486-27044-0

DUBLINERS, James Joyce. Fifteen stories offer vivid, tightly focused observations of the lives of Dublin's poorer classes. At least one, "The Dead," is considered a masterpiece. Reprinted complete and unabridged from standard edition. 160pp. 5$\frac{3}{16}$ x 8¼. 0-486-26870-5

GREAT WEIRD TALES: 14 Stories by Lovecraft, Blackwood, Machen and Others, S. T. Joshi (ed.). 14 spellbinding tales, including "The Sin Eater," by Fiona McLeod, "The Eye Above the Mantel," by Frank Belknap Long, as well as renowned works by R. H. Barlow, Lord Dunsany, Arthur Machen, W. C. Morrow and eight other masters of the genre. 256pp. 5⅜ x 8½. (Available in U.S. only.) 0-486-40436-6

THE BOOK OF THE SACRED MAGIC OF ABRAMELIN THE MAGE, translated by S. MacGregor Mathers. Medieval manuscript of ceremonial magic. Basic document in Aleister Crowley, Golden Dawn groups. 268pp. 5⅜ x 8½. 0-486-23211-5

THE BATTLES THAT CHANGED HISTORY, Fletcher Pratt. Eminent historian profiles 16 crucial conflicts, ancient to modern, that changed the course of civilization. 352pp. 5⅜ x 8½. 0-486-41129-X

NEW RUSSIAN-ENGLISH AND ENGLISH-RUSSIAN DICTIONARY, M. A. O'Brien. This is a remarkably handy Russian dictionary, containing a surprising amount of information, including over 70,000 entries. 366pp. 4½ x 6⅛. 0-486-20208-9

NEW YORK IN THE FORTIES, Andreas Feininger. 162 brilliant photographs by the well-known photographer, formerly with *Life* magazine. Commuters, shoppers, Times Square at night, much else from city at its peak. Captions by John von Hartz. 181pp. 9¼ x 10¾. 0-486-23585-8

INDIAN SIGN LANGUAGE, William Tomkins. Over 525 signs developed by Sioux and other tribes. Written instructions and diagrams. Also 290 pictographs. 111pp. 6⅛ x 9¼. 0-486-22029-X

ANATOMY: A Complete Guide for Artists, Joseph Sheppard. A master of figure drawing shows artists how to render human anatomy convincingly. Over 460 illustrations. 224pp. 8⅜ x 11¼. 0-486-27279-6

MEDIEVAL CALLIGRAPHY: Its History and Technique, Marc Drogin. Spirited history, comprehensive instruction manual covers 13 styles (ca. 4th century through 15th). Excellent photographs; directions for duplicating medieval techniques with modern tools. 224pp. 8⅜ x 11¼. 0-486-26142-5

DRIED FLOWERS: How to Prepare Them, Sarah Whitlock and Martha Rankin. Complete instructions on how to use silica gel, meal and borax, perlite aggregate, sand and borax, glycerine and water to create attractive permanent flower arrangements. 12 illustrations. 32pp. 5⅜ x 8½. 0-486-21802-3

EASY-TO-MAKE BIRD FEEDERS FOR WOODWORKERS, Scott D. Campbell. Detailed, simple-to-use guide for designing, constructing, caring for and using feeders. Text, illustrations for 12 classic and contemporary designs. 96pp. 5⅜ x 8½.
0-486-25847-5

THE COMPLETE BOOK OF BIRDHOUSE CONSTRUCTION FOR WOODWORKERS, Scott D. Campbell. Detailed instructions, illustrations, tables. Also data on bird habitat and instinct patterns. Bibliography. 3 tables. 63 illustrations in 15 figures. 48pp. 5¼ x 8½. 0-486-24407-5

SCOTTISH WONDER TALES FROM MYTH AND LEGEND, Donald A. Mackenzie. 16 lively tales tell of giants rumbling down mountainsides, of a magic wand that turns stone pillars into warriors, of gods and goddesses, evil hags, powerful forces and more. 240pp. 5⅜ x 8½. 0-486-29677-6

THE HISTORY OF UNDERCLOTHES, C. Willett Cunnington and Phyllis Cunnington. Fascinating, well-documented survey covering six centuries of English undergarments, enhanced with over 100 illustrations: 12th-century laced-up bodice, footed long drawers (1795), 19th-century bustles, 19th-century corsets for men, Victorian "bust improvers," much more. 272pp. 5⅜ x 8½. 0-486-27124-2

ARTS AND CRAFTS FURNITURE: The Complete Brooks Catalog of 1912, Brooks Manufacturing Co. Photos and detailed descriptions of more than 150 now very collectible furniture designs from the Arts and Crafts movement depict davenports, settees, buffets, desks, tables, chairs, bedsteads, dressers and more, all built of solid, quarter-sawed oak. Invaluable for students and enthusiasts of antiques, Americana and the decorative arts. 80pp. 6½ x 9¼. 0-486-27471-3

WILBUR AND ORVILLE: A Biography of the Wright Brothers, Fred Howard. Definitive, crisply written study tells the full story of the brothers' lives and work. A vividly written biography, unparalleled in scope and color, that also captures the spirit of an extraordinary era. 560pp. 6⅛ x 9¼. 0-486-40297-5

THE ARTS OF THE SAILOR: Knotting, Splicing and Ropework, Hervey Garrett Smith. Indispensable shipboard reference covers tools, basic knots and useful hitches; handsewing and canvas work, more. Over 100 illustrations. Delightful reading for sea lovers. 256pp. 5⅜ x 8½. 0-486-26440-8

FRANK LLOYD WRIGHT'S FALLINGWATER: The House and Its History, Second, Revised Edition, Donald Hoffmann. A total revision—both in text and illustrations—of the standard document on Fallingwater, the boldest, most personal architectural statement of Wright's mature years, updated with valuable new material from the recently opened Frank Lloyd Wright Archives. "Fascinating"—*The New York Times*. 116 illustrations. 128pp. 9¼ x 10¾. 0-486-27430-6

PHOTOGRAPHIC SKETCHBOOK OF THE CIVIL WAR, Alexander Gardner. 100 photos taken on field during the Civil War. Famous shots of Manassas Harper's Ferry, Lincoln, Richmond, slave pens, etc. 244pp. 10⅝ x 8¼. 0-486-22731-6

FIVE ACRES AND INDEPENDENCE, Maurice G. Kains. Great back-to-the-land classic explains basics of self-sufficient farming. The one book to get. 95 illustrations. 397pp. 5⅜ x 8½. 0-486-20974-1

A MODERN HERBAL, Margaret Grieve. Much the fullest, most exact, most useful compilation of herbal material. Gigantic alphabetical encyclopedia, from aconite to zedoary, gives botanical information, medical properties, folklore, economic uses, much else. Indispensable to serious reader. 161 illustrations. 888pp. 6½ x 9¼. 2-vol. set. (Available in U.S. only.)   Vol. I: 0-486-22798-7   Vol. II: 0-486-22799-5

HIDDEN TREASURE MAZE BOOK, Dave Phillips. Solve 34 challenging mazes accompanied by heroic tales of adventure. Evil dragons, people-eating plants, blood-thirsty giants, many more dangerous adversaries lurk at every twist and turn. 34 mazes, stories, solutions. 48pp. 8¼ x 11.   0-486-24566-7

LETTERS OF W. A. MOZART, Wolfgang A. Mozart. Remarkable letters show bawdy wit, humor, imagination, musical insights, contemporary musical world; includes some letters from Leopold Mozart. 276pp. 5⅜ x 8½.   0-486-22859-2

BASIC PRINCIPLES OF CLASSICAL BALLET, Agrippina Vaganova. Great Russian theoretician, teacher explains methods for teaching classical ballet. 118 illustrations. 175pp. 5⅜ x 8½.   0-486-22036-2

THE JUMPING FROG, Mark Twain. Revenge edition. The original story of The Celebrated Jumping Frog of Calaveras County, a hapless French translation, and Twain's hilarious "retranslation" from the French. 12 illustrations. 66pp. 5⅜ x 8½.
0-486-22686-7

BEST REMEMBERED POEMS, Martin Gardner (ed.). The 126 poems in this superb collection of 19th- and 20th-century British and American verse range from Shelley's "To a Skylark" to the impassioned "Renascence" of Edna St. Vincent Millay and to Edward Lear's whimsical "The Owl and the Pussycat." 224pp. 5⅜ x 8½.
0-486-27165-X

COMPLETE SONNETS, William Shakespeare. Over 150 exquisite poems deal with love, friendship, the tyranny of time, beauty's evanescence, death and other themes in language of remarkable power, precision and beauty. Glossary of archaic terms. 80pp. 5³⁄₁₆ x 8¼.   0-486-26686-9

HISTORIC HOMES OF THE AMERICAN PRESIDENTS, Second, Revised Edition, Irvin Haas. A traveler's guide to American Presidential homes, most open to the public, depicting and describing homes occupied by every American President from George Washington to George Bush. With visiting hours, admission charges, travel routes. 175 photographs. Index. 160pp. 8¼ x 11.   0-486-26751-2

THE WIT AND HUMOR OF OSCAR WILDE, Alvin Redman (ed.). More than 1,000 ripostes, paradoxes, wisecracks: Work is the curse of the drinking classes; I can resist everything except temptation; etc. 258pp. 5⅜ x 8½.   0-486-20602-5

SHAKESPEARE LEXICON AND QUOTATION DICTIONARY, Alexander Schmidt. Full definitions, locations, shades of meaning in every word in plays and poems. More than 50,000 exact quotations. 1,485pp. 6½ x 9¼. 2-vol. set.
Vol. 1: 0-486-22726-X   Vol. 2: 0-486-22727-8

SELECTED POEMS, Emily Dickinson. Over 100 best-known, best-loved poems by one of America's foremost poets, reprinted from authoritative early editions. No comparable edition at this price. Index of first lines. 64pp. 5³⁄₁₆ x 8¼. 0-486-26466-1

THE INSIDIOUS DR. FU-MANCHU, Sax Rohmer. The first of the popular mystery series introduces a pair of English detectives to their archnemesis, the diabolical Dr. Fu-Manchu. Flavorful atmosphere, fast-paced action, and colorful characters enliven this classic of the genre. 208pp. 5³⁄₁₆ x 8¼.   0-486-29898-1

THE MALLEUS MALEFICARUM OF KRAMER AND SPRENGER, translated by Montague Summers. Full text of most important witchhunter's "bible," used by both Catholics and Protestants. 278pp. 6⅝ x 10. 0-486-22802-9

SPANISH STORIES/CUENTOS ESPAÑOLES: A Dual-Language Book, Angel Flores (ed.). Unique format offers 13 great stories in Spanish by Cervantes, Borges, others. Faithful English translations on facing pages. 352pp. 5⅜ x 8½. 0-486-25399-6

GARDEN CITY, LONG ISLAND, IN EARLY PHOTOGRAPHS, 1869–1919, Mildred H. Smith. Handsome treasury of 118 vintage pictures, accompanied by carefully researched captions, document the Garden City Hotel fire (1899), the Vanderbilt Cup Race (1908), the first airmail flight departing from the Nassau Boulevard Aerodrome (1911), and much more. 96pp. 8⅞ x 11¾. 0-486-40669-5

OLD QUEENS, N.Y., IN EARLY PHOTOGRAPHS, Vincent F. Seyfried and William Asadorian. Over 160 rare photographs of Maspeth, Jamaica, Jackson Heights, and other areas. Vintage views of DeWitt Clinton mansion, 1939 World's Fair and more. Captions. 192pp. 8⅞ x 11. 0-486-26358-4

CAPTURED BY THE INDIANS: 15 Firsthand Accounts, 1750-1870, Frederick Drimmer. Astounding true historical accounts of grisly torture, bloody conflicts, relentless pursuits, miraculous escapes and more, by people who lived to tell the tale. 384pp. 5⅜ x 8½. 0-486-24901-8

THE WORLD'S GREAT SPEECHES (Fourth Enlarged Edition), Lewis Copeland, Lawrence W. Lamm, and Stephen J. McKenna. Nearly 300 speeches provide public speakers with a wealth of updated quotes and inspiration–from Pericles' funeral oration and William Jennings Bryan's "Cross of Gold Speech" to Malcolm X's powerful words on the Black Revolution and Earl of Spenser's tribute to his sister, Diana, Princess of Wales. 944pp. 5⅜ x 8⅜. 0-486-40903-1

THE BOOK OF THE SWORD, Sir Richard F. Burton. Great Victorian scholar/adventurer's eloquent, erudite history of the "queen of weapons"–from prehistory to early Roman Empire. Evolution and development of early swords, variations (sabre, broadsword, cutlass, scimitar, etc.), much more. 336pp. 6⅛ x 9¼. 0-486-25434-8

AUTOBIOGRAPHY: The Story of My Experiments with Truth, Mohandas K. Gandhi. Boyhood, legal studies, purification, the growth of the Satyagraha (nonviolent protest) movement. Critical, inspiring work of the man responsible for the freedom of India. 480pp. 5⅜ x 8½. (Available in U.S. only.) 0-486-24593-4

CELTIC MYTHS AND LEGENDS, T. W. Rolleston. Masterful retelling of Irish and Welsh stories and tales. Cuchulain, King Arthur, Deirdre, the Grail, many more. First paperback edition. 58 full-page illustrations. 512pp. 5⅜ x 8½. 0-486-26507-2

THE PRINCIPLES OF PSYCHOLOGY, William James. Famous long course complete, unabridged. Stream of thought, time perception, memory, experimental methods; great work decades ahead of its time. 94 figures. 1,391pp. 5⅜ x 8½. 2-vol. set. Vol. I: 0-486-20381-6 Vol. II: 0-486-20382-4

THE WORLD AS WILL AND REPRESENTATION, Arthur Schopenhauer. Definitive English translation of Schopenhauer's life work, correcting more than 1,000 errors, omissions in earlier translations. Translated by E. F. J. Payne. Total of 1,269pp. 5⅜ x 8½. 2-vol. set. Vol. 1: 0-486-21761-2 Vol. 2: 0-486-21762-0

MAGIC AND MYSTERY IN TIBET, Madame Alexandra David-Neel. Experiences among lamas, magicians, sages, sorcerers, Bonpa wizards. A true psychic discovery. 32 illustrations. 321pp. 5⅜ x 8½. (Available in U.S. only.) 0-486-22682-4

THE EGYPTIAN BOOK OF THE DEAD, E. A. Wallis Budge. Complete reproduction of Ani's papyrus, finest ever found. Full hieroglyphic text, interlinear transliteration, word-for-word translation, smooth translation. 533pp. 6½ x 9¼. 0-486-21866-X

HISTORIC COSTUME IN PICTURES, Braun & Schneider. Over 1,450 costumed figures in clearly detailed engravings–from dawn of civilization to end of 19th century. Captions. Many folk costumes. 256pp. 8⅜ x 11¾. 0-486-23150-X

MATHEMATICS FOR THE NONMATHEMATICIAN, Morris Kline. Detailed, college-level treatment of mathematics in cultural and historical context, with numerous exercises. Recommended Reading Lists. Tables. Numerous figures. 641pp. 5⅜ x 8½. 0-486-24823-2

PROBABILISTIC METHODS IN THE THEORY OF STRUCTURES, Isaac Elishakoff. Well-written introduction covers the elements of the theory of probability from two or more random variables, the reliability of such multivariable structures, the theory of random function, Monte Carlo methods of treating problems incapable of exact solution, and more. Examples. 502pp. 5⅜ x 8½. 0-486-40691-1

THE RIME OF THE ANCIENT MARINER, Gustave Doré, S. T. Coleridge. Doré's finest work; 34 plates capture moods, subtleties of poem. Flawless full-size reproductions printed on facing pages with authoritative text of poem. "Beautiful. Simply beautiful."–*Publisher's Weekly.* 77pp. 9¼ x 12. 0-486-22305-1

SCULPTURE: Principles and Practice, Louis Slobodkin. Step-by-step approach to clay, plaster, metals, stone; classical and modern. 253 drawings, photos. 255pp. 8⅛ x 11. 0-486-22960-2

THE INFLUENCE OF SEA POWER UPON HISTORY, 1660–1783, A. T. Mahan. Influential classic of naval history and tactics still used as text in war colleges. First paperback edition. 4 maps. 24 battle plans. 640pp. 5⅜ x 8½. 0-486-25509-3

THE STORY OF THE TITANIC AS TOLD BY ITS SURVIVORS, Jack Winocour (ed.). What it was really like. Panic, despair, shocking inefficiency, and a little heroism. More thrilling than any fictional account. 26 illustrations. 320pp. 5⅜ x 8½. 0-486-20610-6

ONE TWO THREE . . . INFINITY: Facts and Speculations of Science, George Gamow. Great physicist's fascinating, readable overview of contemporary science: number theory, relativity, fourth dimension, entropy, genes, atomic structure, much more. 128 illustrations. Index. 352pp. 5⅜ x 8½. 0-486-25664-2

DALÍ ON MODERN ART: The Cuckolds of Antiquated Modern Art, Salvador Dalí. Influential painter skewers modern art and its practitioners. Outrageous evaluations of Picasso, Cézanne, Turner, more. 15 renderings of paintings discussed. 44 calligraphic decorations by Dalí. 96pp. 5⅜ x 8½. (Available in U.S. only.) 0-486-29220-7

ANTIQUE PLAYING CARDS: A Pictorial History, Henry René D'Allemagne. Over 900 elaborate, decorative images from rare playing cards (14th–20th centuries): Bacchus, death, dancing dogs, hunting scenes, royal coats of arms, players cheating, much more. 96pp. 9¼ x 12¼. 0-486-29265-7

MAKING FURNITURE MASTERPIECES: 30 Projects with Measured Drawings, Franklin H. Gottshall. Step-by-step instructions, illustrations for constructing handsome, useful pieces, among them a Sheraton desk, Chippendale chair, Spanish desk, Queen Anne table and a William and Mary dressing mirror. 224pp. 8⅛ x 11¼. 0-486-29338-6

NORTH AMERICAN INDIAN DESIGNS FOR ARTISTS AND CRAFTSPEOPLE, Eva Wilson. Over 360 authentic copyright-free designs adapted from Navajo blankets, Hopi pottery, Sioux buffalo hides, more. Geometrics, symbolic figures, plant and animal motifs, etc. 128pp. 8⅜ x 11. (Not for sale in the United Kingdom.) 0-486-25341-4

THE FOSSIL BOOK: A Record of Prehistoric Life, Patricia V. Rich et al. Profusely illustrated definitive guide covers everything from single-celled organisms and dinosaurs to birds and mammals and the interplay between climate and man. Over 1,500 illustrations. 760pp. 7½ x 10⅛. 0-486-29371-8

VICTORIAN ARCHITECTURAL DETAILS: Designs for Over 700 Stairs, Mantels, Doors, Windows, Cornices, Porches, and Other Decorative Elements, A. J. Bicknell & Company. Everything from dormer windows and piazzas to balconies and gable ornaments. Also includes elevations and floor plans for handsome, private residences and commercial structures. 80pp. 9⅜ x 12¼. 0-486-44015-X

WESTERN ISLAMIC ARCHITECTURE: A Concise Introduction, John D. Hoag. Profusely illustrated critical appraisal compares and contrasts Islamic mosques and palaces–from Spain and Egypt to other areas in the Middle East. 139 illustrations. 128pp. 6 x 9. 0-486-43760-4

CHINESE ARCHITECTURE: A Pictorial History, Liang Ssu-ch'eng. More than 240 rare photographs and drawings depict temples, pagodas, tombs, bridges, and imperial palaces comprising much of China's architectural heritage. 152 halftones, 94 diagrams. 232pp. 10¾ x 9⅞. 0-486-43999-2

THE RENAISSANCE: Studies in Art and Poetry, Walter Pater. One of the most talked-about books of the 19th century, *The Renaissance* combines scholarship and philosophy in an innovative work of cultural criticism that examines the achievements of Botticelli, Leonardo, Michelangelo, and other artists. "The holy writ of beauty."–Oscar Wilde. 160pp. 5⅜ x 8½. 0-486-44025-7

A TREATISE ON PAINTING, Leonardo da Vinci. The great Renaissance artist's practical advice on drawing and painting techniques covers anatomy, perspective, composition, light and shadow, and color. A classic of art instruction, it features 48 drawings by Nicholas Poussin and Leon Battista Alberti. 192pp. 5⅜ x 8½. 0-486-44155-5

THE MIND OF LEONARDO DA VINCI, Edward McCurdy. More than just a biography, this classic study by a distinguished historian draws upon Leonardo's extensive writings to offer numerous demonstrations of the Renaissance master's achievements, not only in sculpture and painting, but also in music, engineering, and even experimental aviation. 384pp. 5⅜ x 8½. 0-486-44142-3

WASHINGTON IRVING'S RIP VAN WINKLE, Illustrated by Arthur Rackham. Lovely prints that established artist as a leading illustrator of the time and forever etched into the popular imagination a classic of Catskill lore. 51 full-color plates. 80pp. 8⅜ x 11. 0-486-44242-X

HENSCHE ON PAINTING, John W. Robichaux. Basic painting philosophy and methodology of a great teacher, as expounded in his famous classes and workshops on Cape Cod. 7 illustrations in color on covers. 80pp. 5⅜ x 8½. 0-486-43728-0

LIGHT AND SHADE: A Classic Approach to Three-Dimensional Drawing, Mrs. Mary P. Merrifield. Handy reference clearly demonstrates principles of light and shade by revealing effects of common daylight, sunshine, and candle or artificial light on geometrical solids. 13 plates. 64pp. 5⅜ x 8½. 0-486-44143-1

ASTROLOGY AND ASTRONOMY: A Pictorial Archive of Signs and Symbols, Ernst and Johanna Lehner. Treasure trove of stories, lore, and myth, accompanied by more than 300 rare illustrations of planets, the Milky Way, signs of the zodiac, comets, meteors, and other astronomical phenomena. 192pp. 8⅜ x 11. 0-486-43981-X

JEWELRY MAKING: Techniques for Metal, Tim McCreight. Easy-to-follow instructions and carefully executed illustrations describe tools and techniques, use of gems and enamels, wire inlay, casting, and other topics. 72 line illustrations and diagrams. 176pp. 8¼ x 10⅞. 0-486-44043-5

MAKING BIRDHOUSES: Easy and Advanced Projects, Gladstone Califf. Easy-to-follow instructions include diagrams for everything from a one-room house for bluebirds to a forty-two-room structure for purple martins. 56 plates; 4 figures. 80pp. 8¾ x 6⅜. 0-486-44183-0

LITTLE BOOK OF LOG CABINS: How to Build and Furnish Them, William S. Wicks. Handy how-to manual, with instructions and illustrations for building cabins in the Adirondack style, fireplaces, stairways, furniture, beamed ceilings, and more. 102 line drawings. 96pp. 8¾ x 6⅜. 0-486-44259-4

THE SEASONS OF AMERICA PAST, Eric Sloane. From "sugaring time" and strawberry picking to Indian summer and fall harvest, a whole year's activities described in charming prose and enhanced with 79 of the author's own illustrations. 160pp. 8¼ x 11. 0-486-44220-9

THE METROPOLIS OF TOMORROW, Hugh Ferriss. Generous, prophetic vision of the metropolis of the future, as perceived in 1929. Powerful illustrations of towering structures, wide avenues, and rooftop parks–all features in many of today's modern cities. 59 illustrations. 144pp. 8¼ x 11. 0-486-43727-2

THE PATH TO ROME, Hilaire Belloc. This 1902 memoir abounds in lively vignettes from a vanished time, recounting a pilgrimage on foot across the Alps and Apennines in order to "see all Europe which the Christian Faith has saved." 77 of the author's original line drawings complement his sparkling prose. 272pp. 5⅜ x 8½. 0-486-44001-X

THE HISTORY OF RASSELAS: Prince of Abissinia, Samuel Johnson. Distinguished English writer attacks eighteenth-century optimism and man's unrealistic estimates of what life has to offer. 112pp. 5⅜ x 8½. 0-486-44094-X

A VOYAGE TO ARCTURUS, David Lindsay. A brilliant flight of pure fancy, where wild creatures crowd the fantastic landscape and demented torturers dominate victims with their bizarre mental powers. 272pp. 5⅜ x 8½. 0-486-44198-9

Paperbound unless otherwise indicated. Available at your book dealer, online at **www.doverpublications.com**, or by writing to Dept. GI, Dover Publications, Inc., 31 East 2nd Street, Mineola, NY 11501. For current price information or for free catalogs (please indicate field of interest), write to Dover Publications or log on to **www.doverpublications.com** and see every Dover book in print. Dover publishes more than 500 books each year on science, elementary and advanced mathematics, biology, music, art, literary history, social sciences, and other areas.